How to Build Shoebox FORDS MERCURYS 1949-1954

By Rich Johnson

First published in 1991 by Motorbooks International Publishers & Wholesalers, PO Box 2, 729 Prospect Avenue, Osceola, WI 54020 USA

Created by Tex Smith Publishing Co. PO Box 726, 29 East Wallace, Driggs, ID 83422

The information in this book is true and complete to the best of our knowledge. All recommendations are made without any guarantee on the part of the author or publisher, who also disclaim any liability incurred in connection with the use of this data or specific details

We recognize that some words, model names and designations, for example, mentioned herein are the property of the trademark holder. We use them for identification purposes only. This is not an official publication

Motorbooks International recommends you follow all safety procedures when working on your vehicle. Wear eye protection and respiration filter, especially when painting and around tools. Always dispose of hazardous fluids, batteries, tires and parts properly and safely to protect our environment

Motorbooks International books are also available at discounts in bulk quantity for industrial or sales-promotional use. For details write to Special Sales Manager at the Publisher's address

Library of Congress Cataloging-in-Publication Data

How to build shoe boxes and mercurys / from the editors of Hot rod mechanix
p. cm.
ISBN 0-87938-478-6
1. Ford automobile--Customizing. 2. Mercury automobile--Customizing. I.Hot rod mechanix.
TL215.F7H63 1991
629.28'722--dc20 90-37745
CIP

Printed and bound in the United States of America

CONTENTS

Publisher LEROI TEX SMITH
Editor RICHARD JOHNSON
Tech Editor RON CERIDONO
Art Director BOB REECE
Art Assistant VICKY DAVIDSON
Art Assistant TIM WEEKLEY
Copy Editor BECKY JAYE
Circulation JANET SMITH

FOREWORD

A SHOEBOX IS....

An absolute delight. In the framework of this book, we restrict the term Shoebox (which we sometimes now call a Boxcar) to mean Fords and Mercurys of 1949-'54 vintage. This is the very narrow definition that the term originally implied, but it has now been generally broadened to mean all of the American cars from about 1949-1960. Essentially, a shoebox design is one that is square, not too big, and doesn't seem to fit anywhere else. It is not an Econobox of the late Fifties, early Sixties. It probably doesn't even include the low silhouette MoPars of the later 1950s.

Since the 1949-'54 Ford body style was more boxy than the GM styles, early on we didn't apply the term shoebox to anything but the Ford. We didn't even call a Mercury of the era a box. The Merc body was too round. But you can't talk about a shoebox Ford without including the Mercury. The cars are different, but they are so similar, mechanically, that they need to be considered as one. That's why we have lumped them together in this book. In fact, the Lincoln could probably be included, but we just assume that the average car enthusiast will automatically make the connection.

I've been involved with shoeboxes for years. I enjoy them, but I don't necessarily consider them to be paragons of styling. In a way, the shoebox is only a minor step away from the older buggy-spring type Ford product. It is easy to work on, however, and that is a factor that many rod builders are just beginning to understand. Really important to the shoebox popularity in coming years is the possibility of doing an entire chassis swap, especially when you consider the numbers of 'boxes out there with rotted flooring/frames. Anyway, my involvement with the shoebox goes way back. Back to when they were showroom fresh.

Frankly, I didn't know it was a shoebox Ford. I thought it was our new car. Well, it had only a few thousand miles on the odometer, and it was only a year or so old, and compared to some of the other cars in our family, it was new. It was a 1949 Ford 2-door sedan.

It replaced another sort-of new car. A 1948 Plymouth sedan. My father had bought the Plymouth because he worked at a Plymouth/Chrysler garage. I worked there too, but I was driving a chopped 1934 Ford 3-window coupe. The coupe ran rings around the Plymouth, but that wasn't why the shoebox came to our house. The Plymouth fell into disrepute when it failed to climb some backroad (a better term would be backtrail) deep in the mountain wilderness of central Idaho. The MoPar was short on power at 12,000 feet. Fords of that day weren't. Shortly, the 2-door '49 came to live at our house. (It still couldn't beat the '34!)

The problem with that shoebox was that it owned a flathead Ford V8 engine. In those days, we didn't scorn the flatmotor, but it was prone to some eccentricities. Heating could be a problem on those trips through high desert temperatures. We had a carryover from prior Ford V8 years, in the form of a canvas water bag. This bag (or bags, as the trip may require) was filled with water and dangled from a front bumper or hood ornament. It served occupant and car, whatever was required. There was no air conditioner, since motorists of those years were tough. They could take it. Las Vegas at 120 degrees was nothing. The ill-famed Baker Grade was nothing. When the car overheated, the passengers simply parked and waited for cooler nighttime temperatures. We may have been tough, but we would have killed for a modern A/C!

The 1949 and later Ford flathead engine was better than the earlier 59AB versions. But not by a bunch. It was still a 6-volt electrical system, and when cranking the engine, the starter motor seemed to need 50 volts. Nothing was left for the coil. And the carburetor — every old-time Ford owner knew the technique of hammering the top of the carburetor float bowl. A sticking float needle caused gas to seep out every carburetor crack. It was a ritual, better on the 1949 and later Fords, but not much.

But our "new" Ford was very nice. It looked different than the more boxy earlier Fords. We didn't call them Fat Fords then. We had lots of other endearments, all unprintable. The new Ford was lower. It looked more modern. It seemed to ride better (independent front suspension and semi-elliptic rear springs), and it seemed quieter. In retrospect, it was still a box shape. But a sleeker box. Almost immediately, hot rod and custom fans began to modify the shoebox Ford. They did the 1949 and later Mercs as well.

Around the western states, convertibles got white tops and mild dechroming. Hood and decklid chrome was removed. Sometimes side trim was taken off, sometimes other trim was substituted. Box skirts were a natural for the slab-sided rear fenders. Twin exhausts were expected. Lowering blocks took the back down just enough, and flipper discs covered the front wheels. Wide whites were also expected. If the bumper guards were removed, the impression of low and wide was greater. At least, we thought so. Finally, the really tricked-out custom shoebox Ford or Mercury got a special paint job. Black was popular, so was maroon. Purple and Tahitian Red were strong.

Those were new cars, remember. No radical restoration work was needed. Powertrains were fresh, and so was upholstery. Paint and chrome was new. Creating a personalized shoebox Ford or Mercury in the early Fifties was really a snap.

The exception was when an engine was hopped up. While the typical hop-up might be little more than dual carbs and finned heads, it wasn't uncommon to include a cam (usually 3/4 race), and perhaps even headers with dual exhaust. Better ignition didn't come into play unless the car was going to see some serious racing. Unfortunately, those shoebox FoMoCo transmissions were only marginally better than their predecessors. Actually, they may not have been better at all, perhaps we were just lulled into the thought.

Almost immediately, we started seeing the performance end of hot rodding enter the shoebox Ford/Merc picture. Understand that in the early 1950s, the Mercury had not achieved the near mystic status it enjoys today. There were a few mildly customized Mercs around the country, but there were a lot more (Fords too) that were hot rodded. Little or no attention was paid to changing the body, not even mild dechroming. But power was something else.

By 1950, it was apparent that the flathead was going away. The new short-stroke OHV's coming onboard at GM were obvious powerhouses. The Cadillac and Olds V8s meant instant muscle. Over at Chrysler, the hemi V8 looked very good. Hardly had the 1949 Ford/Mercury hit the street until engine swaps were fair game. The Cadillac was an overnight success. Fordillacs showed up everywhere in the country. I don't ever remember anyone calling something a Mercillac! These swaps weren't difficult. Most any backyard builder could do one, and it wasn't long before speed equipment suppliers were offering special bellhousing adapters to mate OHV's to Ford transmissions.

All of this was fun. But it didn't last very long. At least, not long enough. Although we restrict our shoebox coverage in this book to Fords and Mercs, produced from 1949 through 1954, the actual shoebox legacy runs slightly longer than that. Up to about 1960. The real change in boxcars came with the small block Chevrolet engine of 1955. Overnight, the entire face of hot rodding moved from "then" until "now." No one, it seemed, wanted to work with a flathead engine. Cad, Olds and hemi engines in those earlier Fords and Mercs were instantly replaced by the Chevy V8. Even the 1954 OHV Ford engine began to get attention. Suddenly, the 1949-'54 body style of any marque was set aside.

A few of the boxes remained, but the reasoning for owning one seemed fragile. Why have such a car, which needed an engine swap to be competitive and a top chop to be more like the newer cars in styling, when all that was immediately available at the dealer. The era of the Detroit Muscle Car was beginning.

Now, after all these years, the shoeboxes are back. And back in far greater numbers than ever. But to build a boxcar today is not as easy as "back when." The upholstery is bad, the paint is bad, the body panels need repair or replacement, the chassis needs work, suspension components are worn out, engines are shot. Making a shoebox Ford or Mercury today is time consuming, but not difficult. Anyone can build one, as a custom or a restoration, or a hot rod. But it requires patience and a moderate investment.

The beauty of a shoebox, versus the older fat Ford type hot rod, is that with a 'box you already have most of the suspension problems licked. You start with a good independent front suspension, and you have an open drive rearend that is easy to substitute. The worst chassis problem, that of less-than-excellent brakes, can be solved with a front end disc brake swap. Suspension parts are available from a number of different restoration parts suppliers, and finding body pieces at swap meets is still possible. In fact, finding entire usable parts cars at reasonable prices is common.

An added bonus is the new acceptance of four door sedans and station wagons as viable building projects. Until recently, the only Mercury deemed "perfect" for customizing was the 2-door or the convert. But beautifully detailed and chopped 4-doors have been appearing, and this is now making such body styles desirable among the peer groups.

But there is a fly in the soup. Beware doing a frame front suspension graft until you have checked your state laws. This has been a popular way of getting a more modern IFS with disc brakes. But some states have laws on the subject of frame modifications of this sort. Check it out. Merely adding some different type of front suspension, without cutting the frame off, doesn't seem to be a problem.

There is a tremendous surge of interest in the shoebox Fords and Mercs of the early Fifties. They are fun cars to work with, and they make outstanding personalized vehicles.

This book proves it.

LeRoi Tex Smith
Publisher

by John Lee

SHOEBOXES ARE WHERE YOU FIND 'EM

You won't find the kind of shoebox we're talking about in this book among the lint balls under the bed! We're concerned with shoeboxes of the Ford and Mercury persuasion, specifically those squarish types built from 1949 through 1954.

Checking the new *Standard Catalog of Ford, 1903-1990* (Krause Publications, Iola, Wisconsin 54990, $19.95), we find the following total production figures for these years:

	FORD	MERCURY
1949	1,118,740	301,307
1950	1,206,739	293,658
1951	1,013,381	310,397
1952	671,733	172,087
1953	1,244,540	305,863
1954	1,162,743	259,305

So, where are they all now? Most of them are gone, junked out, turned back to scrap metal and shaped into some other product. Some are restored and maintained in enthusiasts' garages, as are others that have been customized or turned into race cars. A few are still being driven on the streets.

And the rest of the survivors are out there somewhere waiting to be discovered. You'll find them wherever you find any old cars — in junkyards, especially those that have been inactive in later years. You'll sometimes find them in backyards and farmyards, and even lining the banks of ditches and streams. So get out there and look. See what you can find.

To get you started, here are a few photos of shoebox Fords and Mercs we've spotted in recent times.

This '49 Merc in an old Colorado salvage yard was missing some parts, but the coupe body was in good shape.

Below-1951 Crestliner was spotted parked beside a street in a small Nebraska town.

Another Colorado salvage yard was home to this '53 Ford Courier a few years ago.

Below-Seen in a South Dakota salvage yard was this '51 Mercury Monterey, a rare piece in fairly good shape.

The same South Dakota yard held this '51 Victoria, but supposedly neither it nor the Monterey were for sale.

An old conversion of a '49 Mercury into a pickup with a '53 Chevy grille ended up in a Kansas salvage yard.

This 1950 Ford sedan has been part of this rustic scene in a Nebraska town for some years.

This chopped '51 Ford appeared in a mid-1950s Rod & Custom spread about a Kansas custom show. On a Salina car lot in the early '80s, it appeared to be very restorable.

1950 Ford Custom Tudor was recently photographed sitting in a yard.

Below-A collector had this '51 sedan sitting in the weeds, literally.

1954 Victoria was photographed at the same place.

We first saw this sectioned '51 convertible in 1958 in just about this condition. There's no evidence it was ever a finished job. Hammer welded seam where sectioning was done is not even visible. Current owner isn't interested in selling.

Dogs prevented us from getting a close-up look at this '52 Mercury. Looks decent, though.

This '51 and the body of a '52-'53 Ranch Wagon were spotted in a storage compound in Missouri.

by John Lee

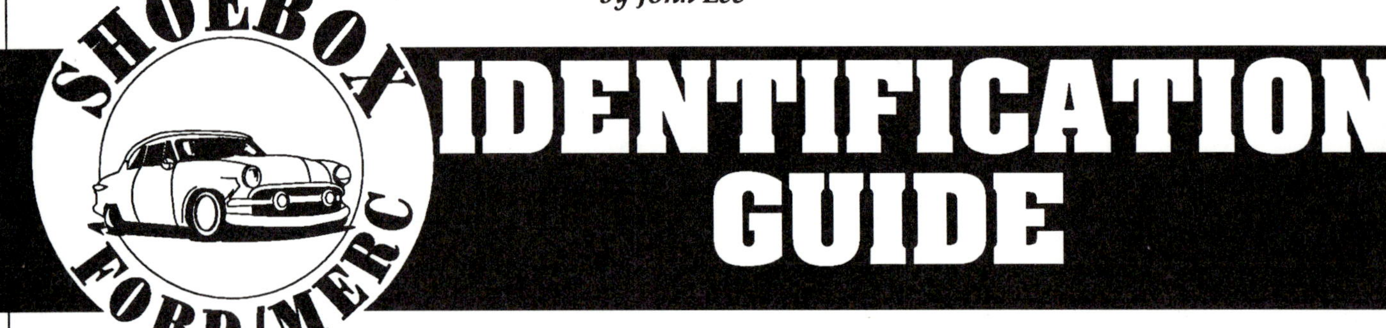

IDENTIFICATION GUIDE

Identification of shoebox Fords and Mercs can sometimes cause confusion among inexperienced enthusiasts who have not yet learned the small details which distinguish between the various years or models. But once a few key identifiers are pointed out, it is quite easy to spot the differences between '49, '50 and '51 Fords, or the distinguishing features of Mercurys of the same era.

For shoebox Fords and Mercs built during the period of 1952-'54, the identification game starts all over again. New body styles were introduced. And while the earlier Fords and Mercs looked absolutely nothing alike, there was some sharing of features between both cars in the '52-'54 period. While it's hardly possible to mistake a '52-'54 Ford for a Mercury of the same era, it is sometimes not so easy to know one year from another during this period.

When it comes to make, model and year identification, the following photographs and captions should be helpful, and downright fun to look at.

Parking lights were an extension of the main grille bar on the '49 Ford. F-O-R-D was spelled out over rounded center section of the top bar. This mild custom has a bull nose strip replacing the hood ornament and has bumper guards removed.

The '49 had pull-type door handles, exposed gas filler cap, propeller-like chrome deck lid handle, deck hinges with a hump. Flared skirts are aftermarket.

Grille bar extension wrapped around corner of the fender, and parking lights were in a separate panel below for 1950.

Below-Both '49 and '50 Ford had the number of cylinders — 6 or 8 — stamped in the center grille bullet.

Ford crest replaced individual letters of '50 hood, and hood ornament was slightly revised.

CUSTOM, embossed in front and side trim piece, is above the trim line on 1950 models, below it on 1949 cars.

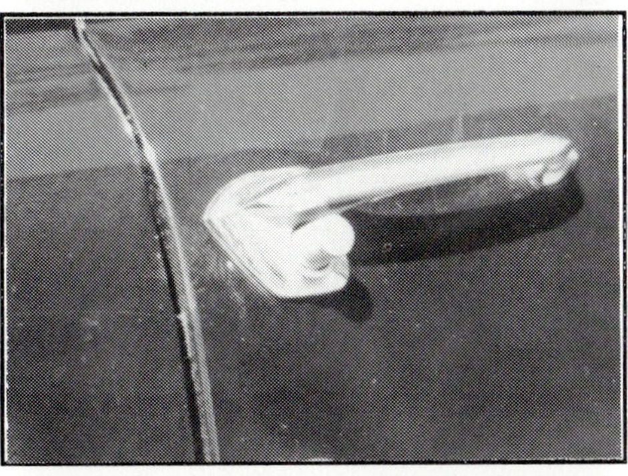

Switch was made to push-button door latches for '50.

The owner of this '50 convertible has added some mild custom touches — chromed license housing/deck handle, chrome windsplits for a '51 look, Crestliner full wheel covers. Triple rib skirts were a Ford accessory for '49, '50 and '51.

The '50 license housing was an inverted U, painted the body color, with crest above. Deck lid hinges were smoother.

Redesigned '51 grille replaced the center spinner with smaller ones set halfway to the sides. Headlight rims are slightly tunneled. There's an airplane hood ornament, redesigned bumpers and round parking light lenses.

Above-1951 Ford Custom rear view was changed with wider taillights, accented by chrome windsplits, trimmer deck lid handle, bright trim below the rear window and redesigned bumpers and guards. Side trim now continued to envelope the rearend. Gas filler was hidden behind a door in '50 and '51.

This '51 Custom Tudor features after-market flared skirts and Fiesta-style spinner wheel covers. Emblem on the back of the front fender identified V8 models in '51.

Mildly customized '51 ragtop is typical of the period. Duals, continental kit, flared skirts, and spotlights are all bolt-ons.

Ford's new styling for '52 featured a full-width grille shell and bar, tunneled headlight rims, windsplit bulges on the rear quarters and jet-like round taillights.

Facelift for '53 used a single grille bar with a center spinner horizontal spear on the rear quarter and revised taillights and ornaments. This Victoria has accessory skirts and grille and tail guards.

Above-Ford introduced the "falsie" continental tire on the Indy pace car in '53, and it became a popular accessory.

1953 taillights had more the look of a jet exhaust. Deck lid handle looks like wings with emblem above. This convertible has '55 wheel covers.

Grille was once again redesigned for '54, still featuring a center spinner flanked by round parking lights. Headlight rims had a slight shade on top.

Taillights were again changed with the addition of three chrome fins in the lens and black "dashes" around the rim.

Full length side spear, with a slight dip ahead of the rear quarter bulge, gave the Customline and Crestline series a longer look.

Rear quarter trim used smaller gravel guard with model script near the top of the bulge. These stainless skirts are aftermarket items.

Popular new model for '54 was the "glasstop," with a tinted clear plexiglass window above the front seat.

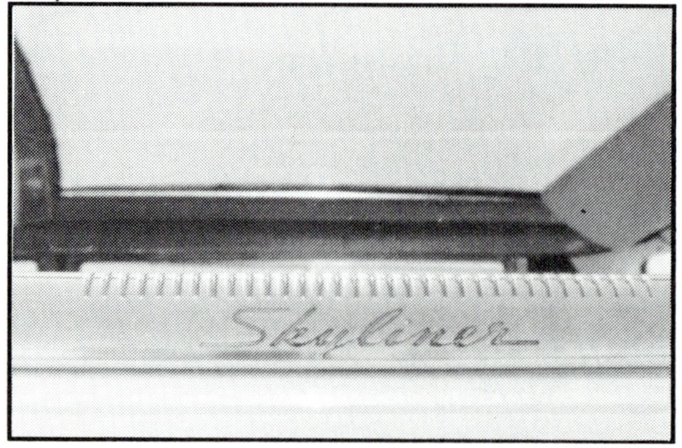

Names of top-of-the-line sport models were stamped into the door sill stainless trim. Skyliner was the name of the new glasstop coupe.
Below-Crestline was the top series for '54, embracing all the sport models plus a sedan and a wagon.

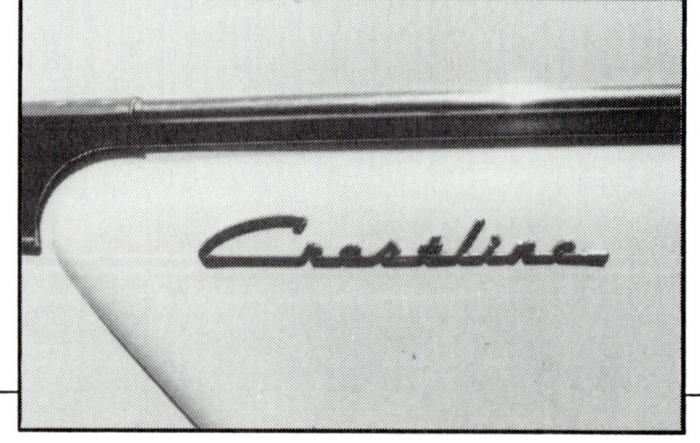

Clean, rounded styling of the 1949 Mercury was originally intended for the Ford. Waterfall-type grille kept heritage of the '46-'48 Merc. Name is spelled out in individual letters on '49, with small, round parking lights, flush headlight rims, and MERCURY stamped into front of side spear.

Side character line with prominent dip in the front door gave '49 Merc its distinction. This is the coupe; there was no Tudor sedan. Horizontal bar deck lid handle and pull-type door handles characterized the '49. Three-piece rear window in this car has been changed to the one-piece '50 type.

For '50, the Merc got a new trim piece on the hood lip with MERCURY stamped into it, large chrome bezels around the parking light and a slightly different side spear with wider nameplate section on the front fender. This convertible has '51 headlight rims.

Without a 2-door hardtop to match the competition, Mercury built a fancy vinyl-top coupe called the Monterey. Script on the door identifies it. This example also has the correct flush headlight rims.

Mercury switched to pushbutton door handles and one-piece rear window for '50, and the deck handle/emblem was a large semi-circle. Skirts were standard on Monterey, optional on others.

Handsome wood-paneled Tudor station wagon body served both Ford and Mercury from '49 through '51, but numbers were small. This is a '51 with accessory sun visor and surfboard (not available on the Ford option list!).

Mercury was back to sharing its body shell with Ford for '52, instead of the small Lincoln as in '49-'51. Carried over with only trim changes for '53, it featured a bumper/grille combination with bullet bumper guards and chrome teeth on the top pan, a chrome-trimmed simulated hood scoop and tunneled headlight rims with chrome inserts.

1953 Mercury side spear ran full length, and small windsplit trim was added to rear quarter bulges. Tower taillights were the same in '52 and '53, deck lid handle had horizontal wings. Merc offered skirts optionally, but these stainless steel items are aftermarket.

Above-Large wrap-around taillight/backup lights and a revised rear bumper changed the rear view. This stock '54 Mercury hardtop featured a continental kit.

Along with Ford, Mercury got a glasstop coupe model for '54, the Sun Valley. Round emblem near the front of the side spear, and vertical indentations painted black in the top bumper bar (for a toothed effect), were changes for '54. This Sun Valley has accessory stainless skirts, rear mount antenna, continental kit, lakes pipes, spots, and chrome headlight rims.

photos by Ryan Johnson

ACCESSORIES

Personalizing a car has always been one of the preferred pastimes for enthusiastic automobile owners. The factory was quick to catch onto this, and during the shoebox years there were lots of factory/dealer accessories available to help an owner express a bit of his own personality in the way his car was equipped.

What the factory or dealership didn't offer, enterprising aftermarket suppliers did. By making the most of the factory/dealer accessory inventory, or by thumbing through such masterpieces of automotive accessory literature as the J.C. Whitney catalog, one could be assured of the opportunity to distinguish his car from all the rest.

Following are photographs of just a few of the many shoebox Ford accessories that were popular during the early Fifties. With the resurgence of interest in these cars, there is a similar popularity for authentic period accessories. If you are lucky enough to find some of these pieces, expect them to command top dollar.

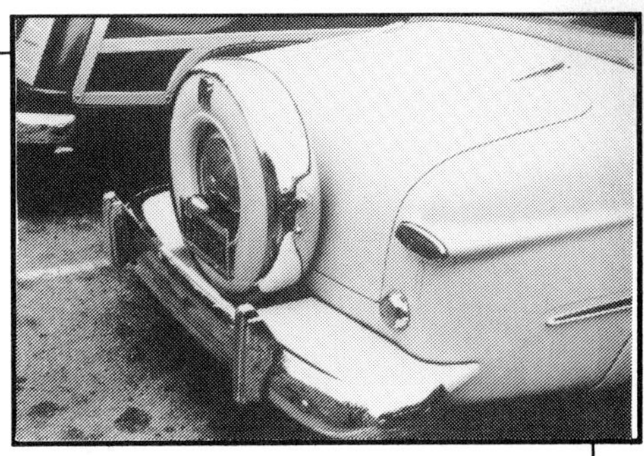

DESERT
WATER BAG

CUSTOM

Ford

by John Lee

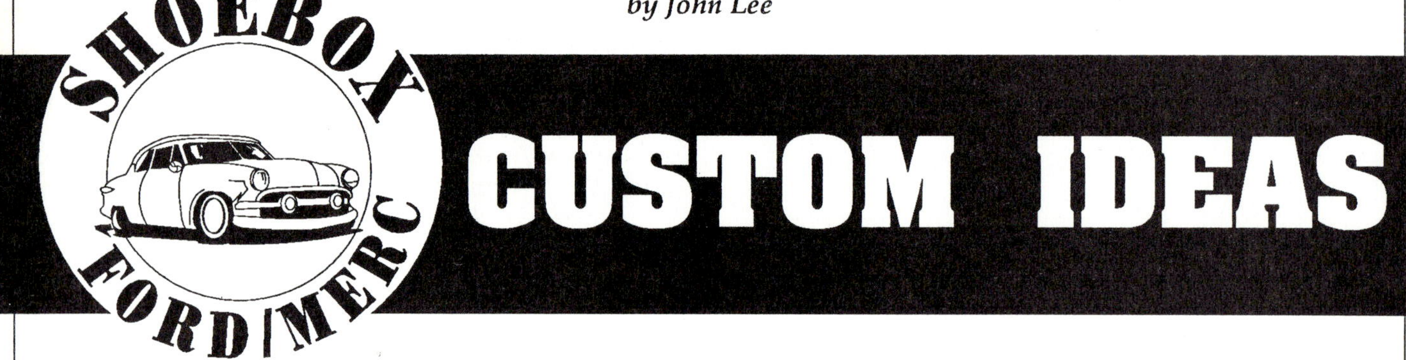

CUSTOM IDEAS

Certainly, among the most memorable customs ever built, shoebox Fords and Mercs rank right up there with the best. These are nostalgia cars, and for anyone who was on the scene in the "good old days," these customs will evoke a lot of memories and stir up old desires. Memories of cars once owned; desires to build a car like this now, because it was somehow beyond reach back then.

That's exactly what this section of the book is supposed to do. It's supposed to take you back to simpler times, provoke memories long since asleep, and reawaken those old desires. Maybe it will even inspire you to get started on your own custom project. If so, we have succeeded.

Many shoebox Fords were mildly customized like this convertible, which has Ford accessory skirts, bullnose molding, bumper guards removed, and '55 Mercury spinner wheel covers.

This '50 Ford Tudor was given the custom treatment with rolled pans, '53 Chevy grille bar and teeth in a molded shell, rounded hood corners, shaved chrome, etc. It was built by Kirby Neal.

Art Bedan still has the '50 convertible he built in the late '50s. It has a trimmed-down '54 Ford grille bar in a '50 Merc shell, headlights frenched with '52 Merc rims, shortened side trim, electric doors, and frenched stock taillight lenses.

Bedan's flathead engine is mildly reworked with a cam and 3-carb manifold.

Top chop on this shoebox features canted B-pillars and one-piece glass in front and rear windows. Grille is a narrowed '55 Plymouth unit. Buick-style two-toning is all in paint.

Along with a top chop, this '50 coupe has been sectioned and fitted with inverted '56 Plymouth side trim for a low look.

Club coupe is devoid of chrome trim and has a molded Merc grille shell with what appears to be a '56 Chrysler bar.

Sectioning and fitting a Merc grille shell leaves an abbreviated opening for the '54 Pontiac grille bar. Hood is filled, with corners rounded.

Little bodywork, but much thought in laying out the light blue scallops on black, created this '50 custom. A straight grille bar with teeth below, and '56 Chevy vertical side trim piece are unique touches.

Ford of Canada put a Merc grille in a Ford and called it Meteor. This one features dechroming and frenched lights.

Bob McCormick's '50 Ford has survived intact since being built in California for Burt Hamrol in 1958. Raising smooth Olds bumpers front and rear makes it look sectioned, and radiused wheel openings give a light look.

Ed Guffey's '50 custom was designed by Guffey and Doug Thompson in the '80s to look as if it could have been built in the '50s. Pontiac-style scoops over headlights, '54 Imperial grille, and molded "portholes" in conjunction with Crestline-type side styling are highlights.

Rear of Guffey's Ford, like front, has '58 Chevy bumper center molded to '51 Ford ends. Exhaust ports are incorporated in rear.

Contemporary style '51 Ford, built by Bob Sipes, features this engine compartment with intricately louvered aluminum panels surrounding small block engine.

Unfinished '52 Ford has a '50 Olds 98 grille bar floating in a Merc shell.

Radical '52-'54 Ford custom has a deep top chop with canted B-pillars, full dechroming, tunneled quad lights, rolled pans and multi-bullet grille. Rear is extended with '56 Packard Clipper light units, flush bubble skirts and a molded housing for license and exhaust tips.

Painted bumper and horizontal tube grille give '52 a wide, low look. Hood corners are rounded.

If Ford had brought out the Ranchero in '52, it would have looked like this. Neat conversion appears to use combination of station wagon and sedan panels.

Built in the '50s, this clean '51 convertible features a straight bar grille, station wagon bumper guards and '57 Chevy side spear, to create two-tone accent.

Flathead in '51 custom was typical of the era, with Edelbrock heads, triple carbs and lots of chrome.

Dean Bordner's sectioned club coupe has survived from the '50s. Corvette grille, narrow opening, and rolled pan give it a clean look.

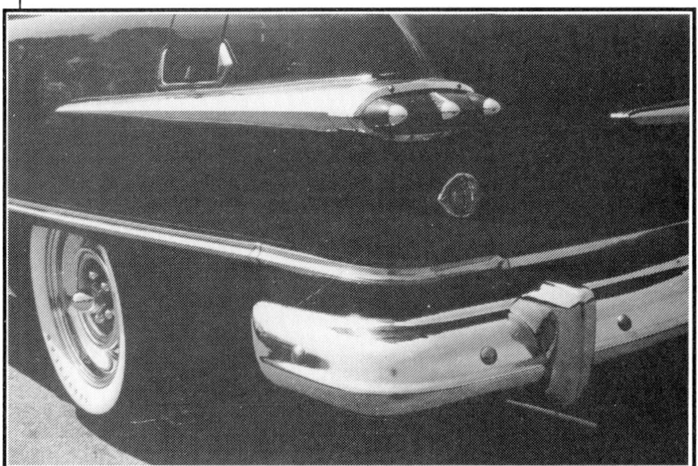

Small bullets added to taillight lenses, and spider wheel trim add to the appeal of this '51.

Rare '54 Ford Courier sedan delivery is updated with a later engine and Skyliner glass roof. Plexiglass hood insert was offered by Ford in '54, so dealers could show off the new OHV V8 engine.

Customizers put Caddy fins on everything in the early '50s. They don't go as well on a shoebox Ford as on some models, but top chop and side trim treatment are neat.

1954 ragtop benefits from a '54 DeSoto grille bar in stock cavity, frenched headlights, solenoid doors, and Merc teeth on rear quarter.

Bumper of this '54 appears to be inverted, to give a rolled pan effect. Grille is '53 Dodge. Note shortened trim, hood and rear quarter scoops, and scallops on roof.

Full-dress flathead graces this Merc engine compartment, with gauges to help tuning.

Along with chopped top, this '54 has molded grille shell with stock bar shortened and "floated" sans center spinner. Rear quarters have scoops.

Both '51-'53 (left) and '54-'55 DeSoto grilles are popular in Merc shells.

A classic custom combination for a Merc is '54 Buick headlight housings, Corvette grille, and Kaiser bumper guard.

The contemporary look was given to this bathtub Merc with vertical quad lights behind translucent lenses, molded pan and single grille bar.

1950 ragtop is cleaned up with shaved hood, deck and door handles, as well as scoops in sides of hood.

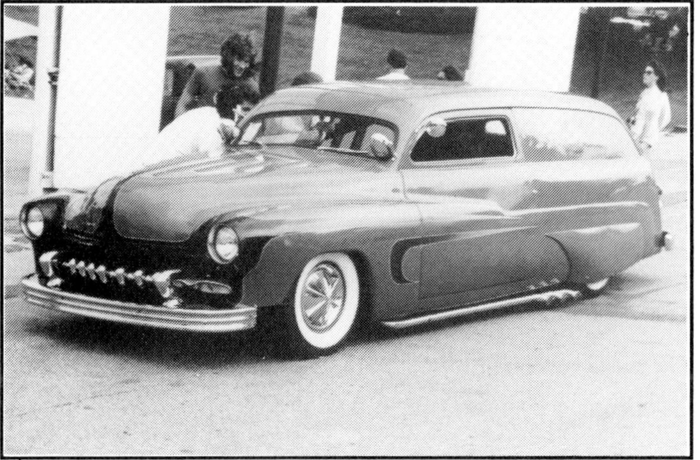

Mercury didn't offer a sedan delivery, but a creative customizer created one of his own. It would have sold!

David Wolk built his '50 with a blend of old and new tricks, including Corvette grille, complete dechroming, and chrome reversed wheels wrapped in black radials.

Merc's side line dip is a natural for custom two-toning, but it hasn't been done often.

A seldom-used top chop approach calls for canting the B-pillar and filling the quarter windows, for a formal look.

Red flames over cream lend a hot rod look to a '49 Merc coupe.

The reverse treatment, light flames on dark background, is demonstrated on this similar Merc.

Two-tone treatment on his Merc is unique, as are exposed header pipes. Grille is a shortened '55 DeSoto item.

Maybe the wildest custom Merc ever built is this one, done in the '60s by Joe Bailon, and restored by him in the '80s for Sterling and Kathy Ashby.

A pair to draw to — full custom '50 Ford coupe and mild '51 Merc convertible.

The '49 coupe James Dean drove in "Rebel Without a Cause" has inspired dozens of custom Mercs. The actual car was mild, with shaved hood and deck, '51 skirts and pencil tips.

Asymmetric treatment has been used with success on this '50 Merc with '62 Merc taillights set into fenders.

Bob Johnson built this '50 "So Fine," featuring a 3-inch top chop, '54 Pontiac and Chevy grille combination. Retaining the side spear adds to the long, low look.

Filled quarter windows and canted B-pillar with top chop give this '50 a formal look.

Ed Lepold's '50 has classic styling, with Buick side spear, '53 DeSoto grille and bumpers, white pearl with gold trim.

Black Naugahyde in 2-inch pleats creates a classic interior in Bob Oman's '50.

Ronnie House is the long-time owner of this mild '50 convertible. Reversing taillights left-to-right, for a cat-eye effect, was a popular trick.

Bob Oman's black '50 coupe is chopped four inches and dechromed, with additional vertical bars replacing the bar in the center of the grille. Hood is shaved, with rounded corners.

Rear of Oman's Merc has '50 Lincoln taillights set in, and pencil tip exhausts.

A pair of classic customs, '49 Merc in foreground and '51 Ford in back, both chopped.

Scallop treatment in white and gray primer is wild. Bumper appears to have been turned upside down, with pan built out over it.

Inverting a second grille shell top section on the bottom yields this oval opening, which is neatly filled with a Corvette bar.

Tunneling lenses into a molded opening is the cleanest treatment with '59 Cadillac lights.

Roger Ward built this clean contemporary '51 Merc, with a top chop and Star wires in black radials. Some trim is ditched, but grille, headlights and taillights were left stock.

Of course, the all-time most famous custom Merc, owing to its many magazine and movie appearances, was that of Bob Hirohata, built by Barris. Doug Thompson built this exact replica for Jack Walker, who has displayed it from coast to coast.

Probably the most radical '51 was this 1984 King of Mercs, by Dave Stuckey. It is chopped and sectioned, with Pontiac bumpers, Chevy grille teeth, custom '54 Merc taillights and Buick power.

Glenn Steck won the Harry Bradley Award at the '89 Leadsled Spectacular, with this '51 convertible. It features a grille bar custom-built from bumper guards, DeSoto bumpers, rear quarter scoops, and hand-formed taillight lenses.

Some Merc owners feel it necessary to remind viewers what their cars are not.

1951 coupe has been lowered and cleaned up with dechroming.

Gary Wakefield's '51 coupe is one of the nicer mild Mercs around, with '54 Chevy grille worked into stock grille ends, frenched and tunneled headlights, shaved hood, deck and door handles, tunneled antenna and burgundy scallops over red finish.

Bill Moore used '69 Lincoln taillights in his '52 Merc, along with custom-built flush skirts and tunneled antenna. Front was cleaned up with removal of bumper guards, filled hood scoop, and Cordoba headlight doors.

Top treatment on Moore's '52 Merc was first to chop, then cover the top with heavily-padded vinyl, and a small rear window opening, for a convertible look.

Some of the changes on this Merc include deep top chop and smaller rear window, Buick side spear, bullets in a '55 Buick grille, cruiser skirts and '56 Packard taillights.

Nice '53 Merc lines are improved with dechroming. Hood trim, grille teeth, and upright guards, door handles and deck trim have been eliminated. Bottom bumper bar was painted to look like a rolled pan.

Above-Rear fender caps have been built for this '53, then Caddy lenses tunneled deeply into them.

Treatment on this '53 Merc is to retain bumper "bombs" but scrap upright sections, then paint the lower pan body color.

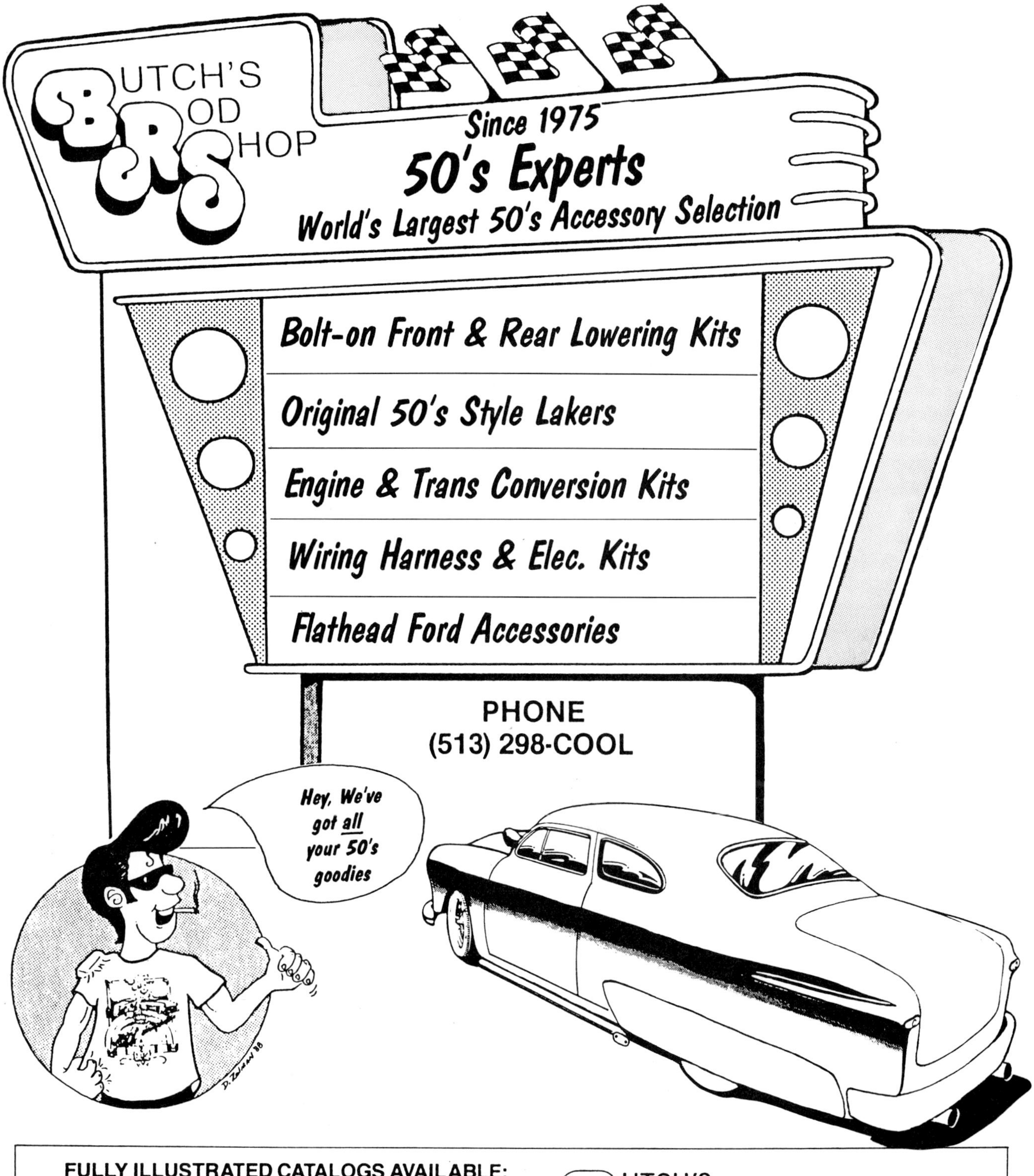
BUTCH'S ROD SHOP
Since 1975
50's Experts
World's Largest 50's Accessory Selection
Bolt-on Front & Rear Lowering Kits
Original 50's Style Lakers
Engine & Trans Conversion Kits
Wiring Harness & Elec. Kits
Flathead Ford Accessories
PHONE
(513) 298-COOL
Hey, We've got all your 50's goodies

by Rich Johnson

FRAMES

Elsewhere in this book, we discuss the hows and whys of swapping a late-model frame under a shoebox Ford or Merc. Of course, not everybody wants to go that route. Actually, there is absolutely nothing wrong with sticking with the stock frame, unless you want a quick way to get the benefits of front disc brakes and late-model suspension and steering.

So, for those who have determined to use the original frame beneath their shoebox, we have gathered up factory diagrams and dimensions for all of the Ford and Mercury frames built during the period of 1949 through 1954.

There are very minor differences between the various years of some of these frames. Most often, it's nothing more than the distance between holes in the frame, or the gauge of steel the units were made from. If you were wandering a wrecking yard looking for a frame to use under a project car, it would be handy to be able to refer to these diagrams when figuring out the differences between a '49/'50 and a '51 Ford frame, for example. This information is particularly important to a restorer who is interested in building a totally authentic automobile. But beyond that, it is also good to have access to all the correct dimensions, so you can check your frame for square or restore damaged areas.

So, have fun studying these diagrams and seeing how many differences you can find between the various models and years

1949-1950 FORD CLOSED CAR

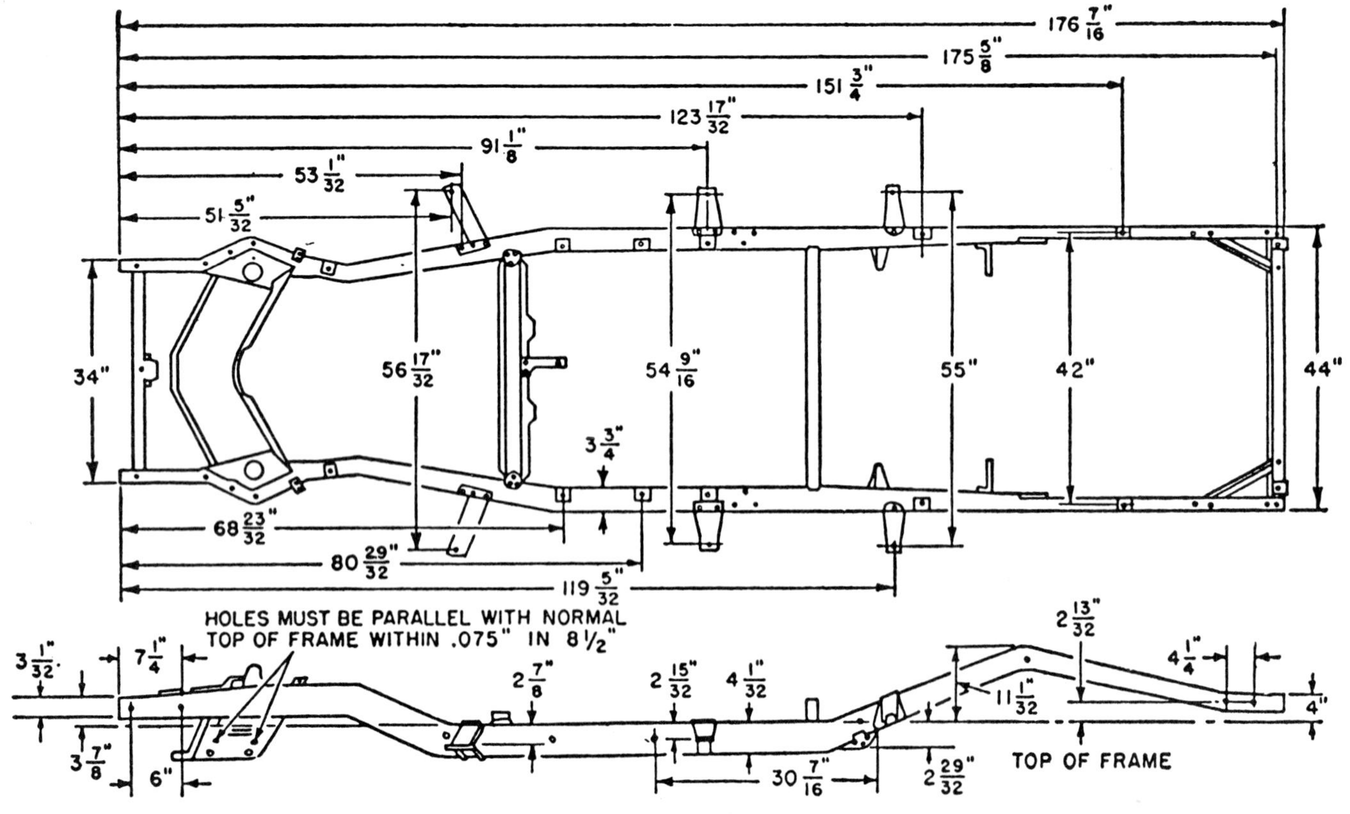

HOLES MUST BE PARALLEL WITH NORMAL TOP OF FRAME WITHIN .075" IN 8 1/2"

TOP OF FRAME

1949-1950 FORD

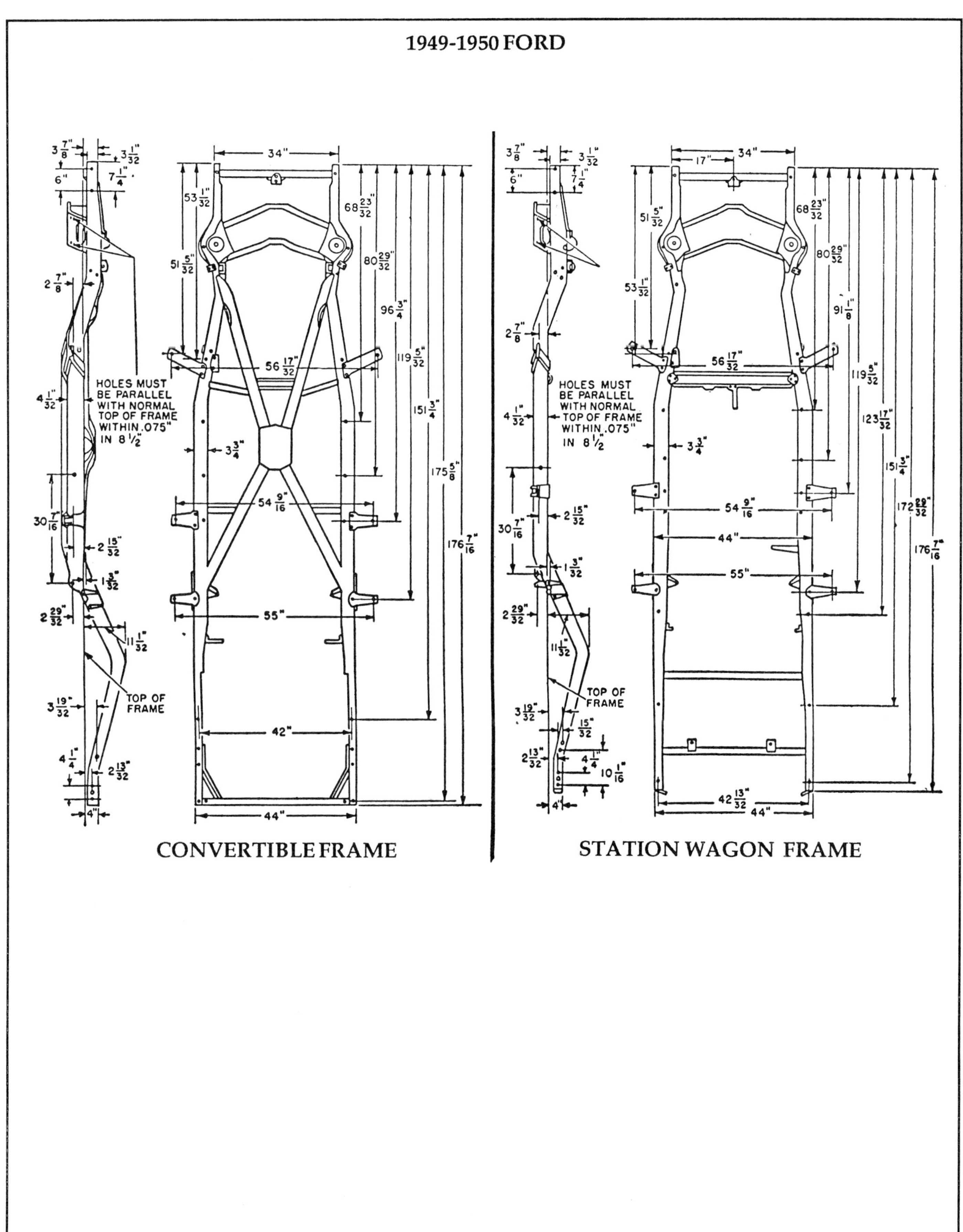

CONVERTIBLE FRAME

STATION WAGON FRAME

1951 FORD CONVERTIBLE

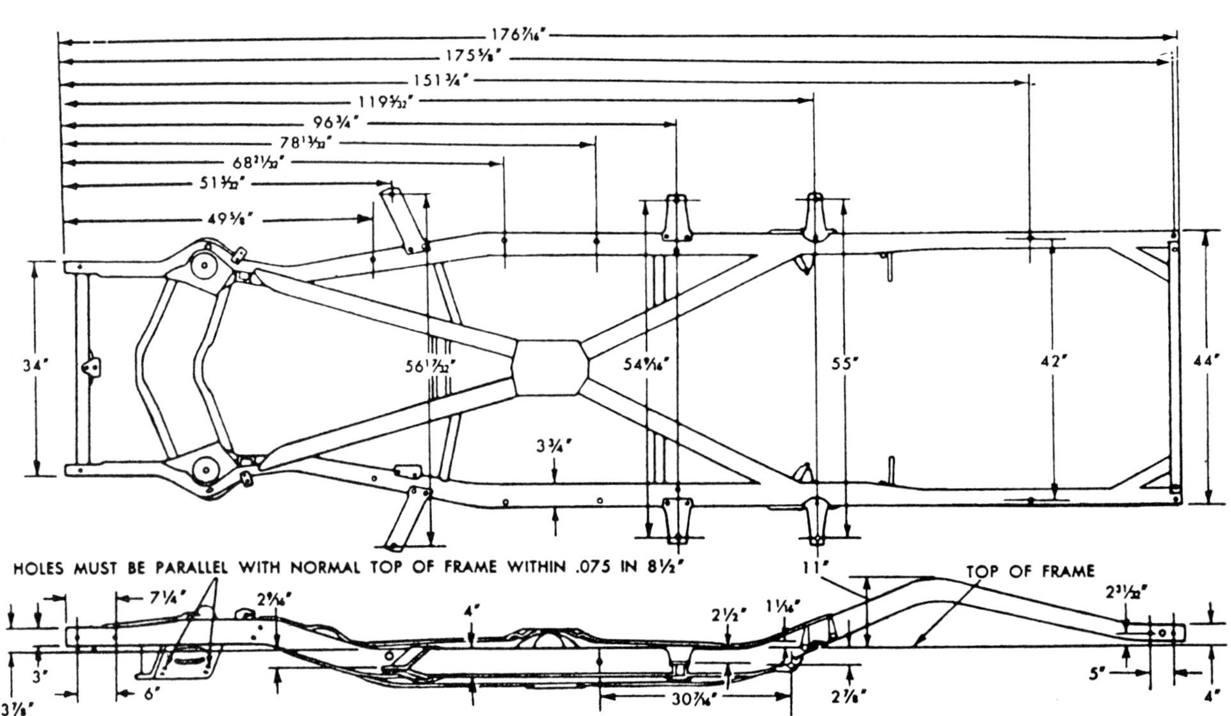

1951 FORD STATION WAGON

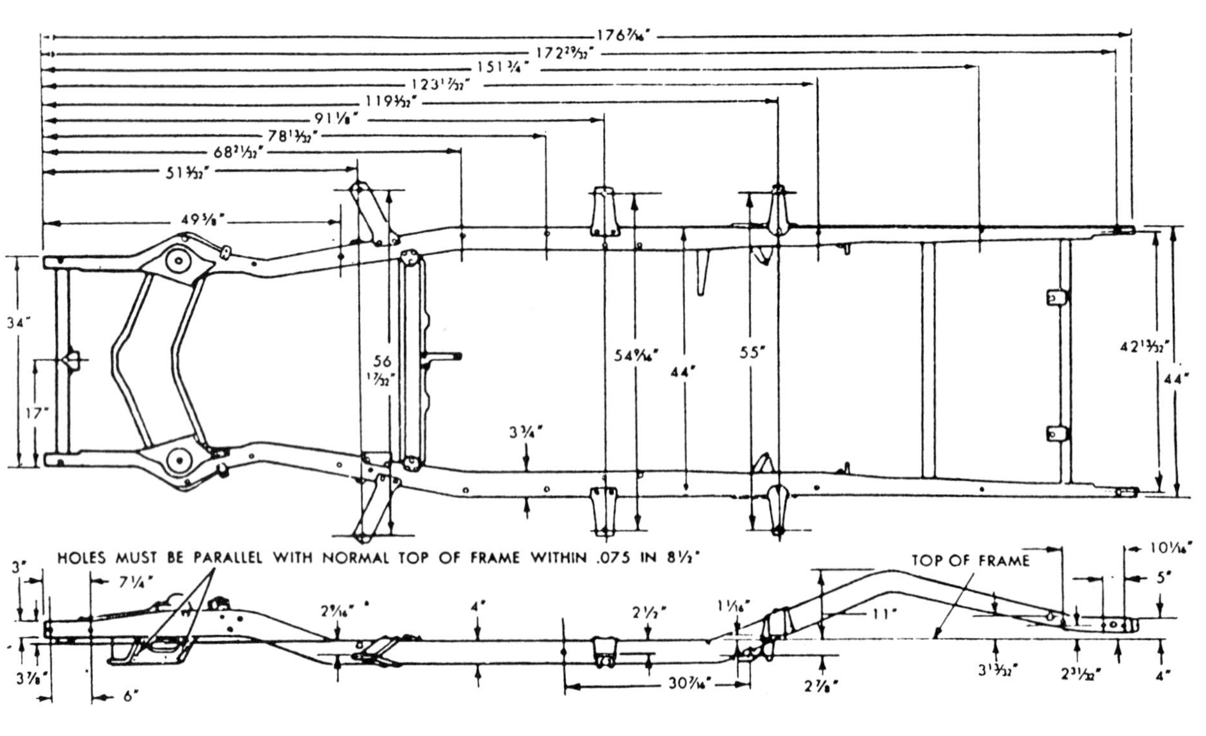

1951 FORD CLOSED CAR

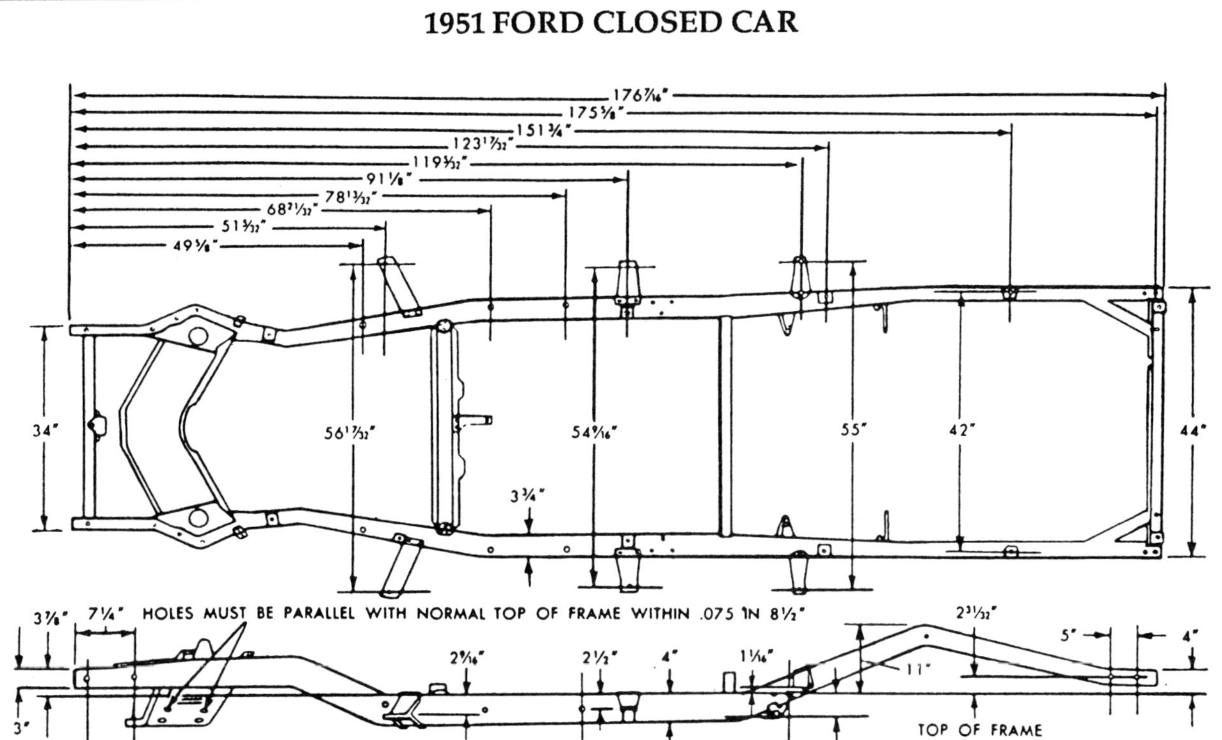

1952-1953 FORD VICTORIA

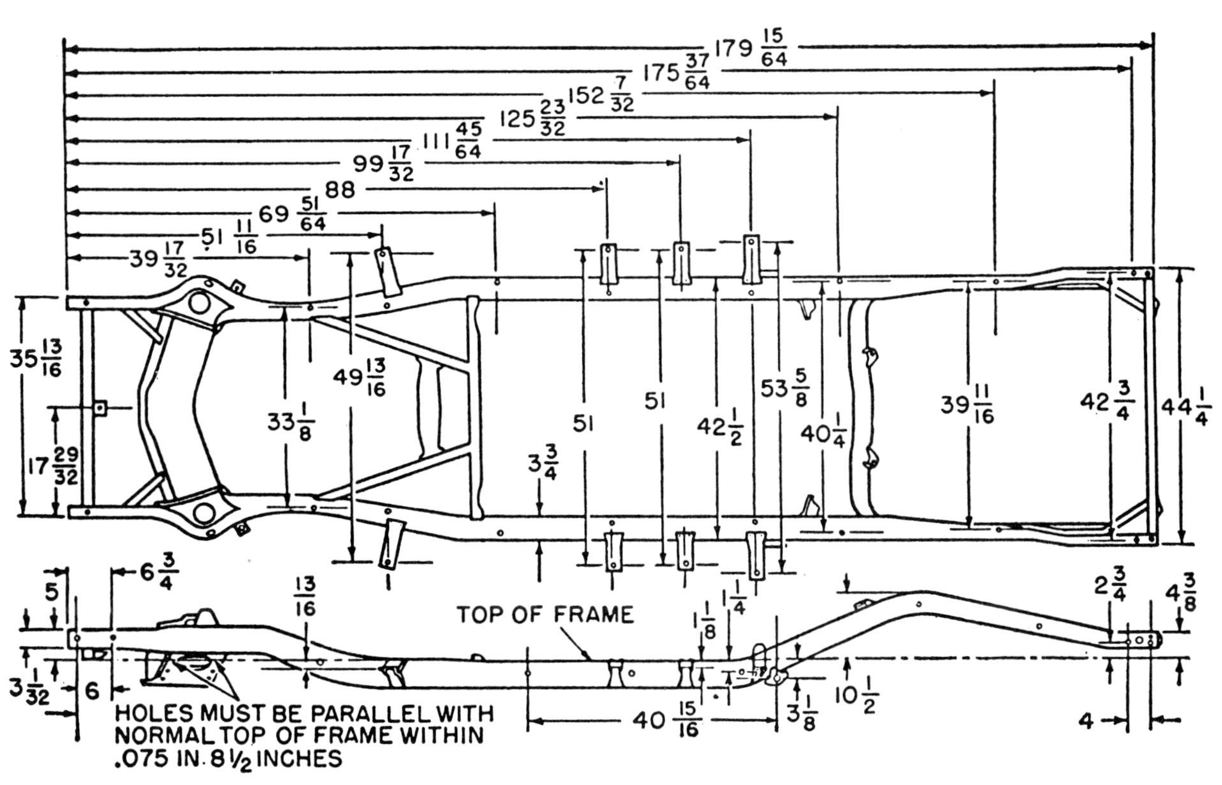

1952-1953 FORD 4 DOOR - 2 DOOR & COUPE

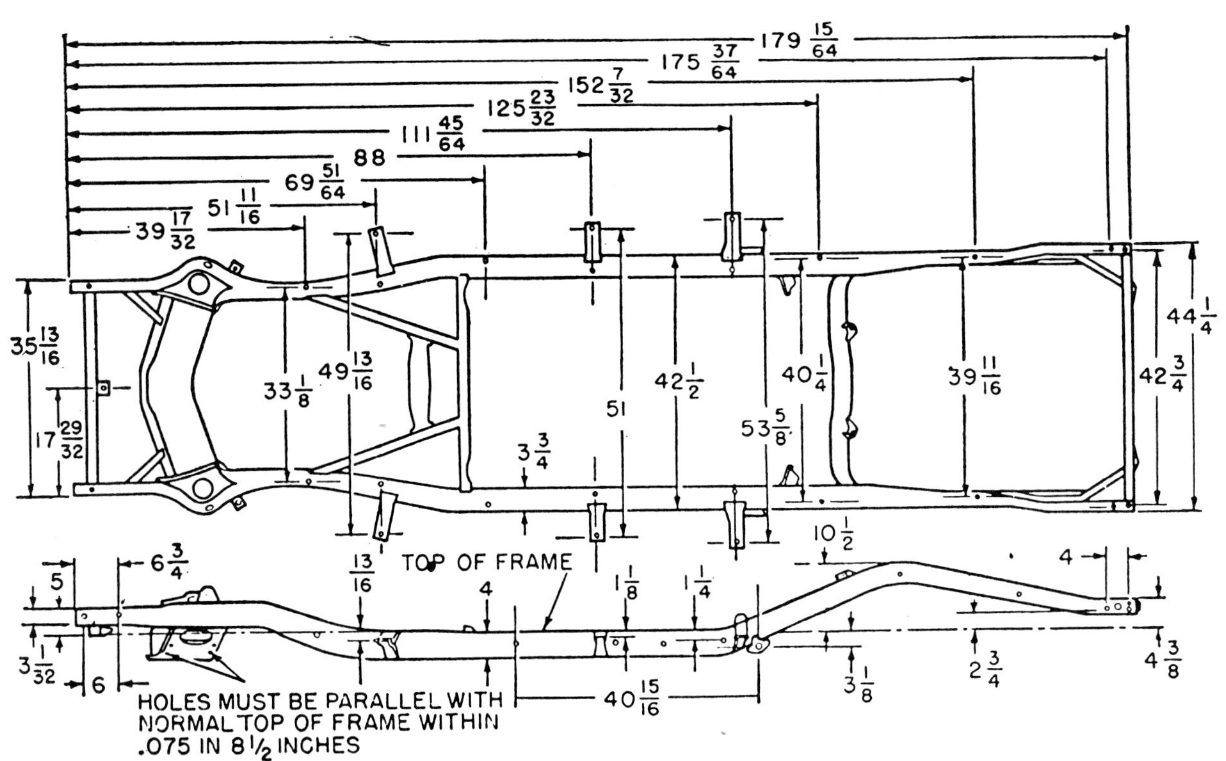

1952-1953 FORD CONVERT

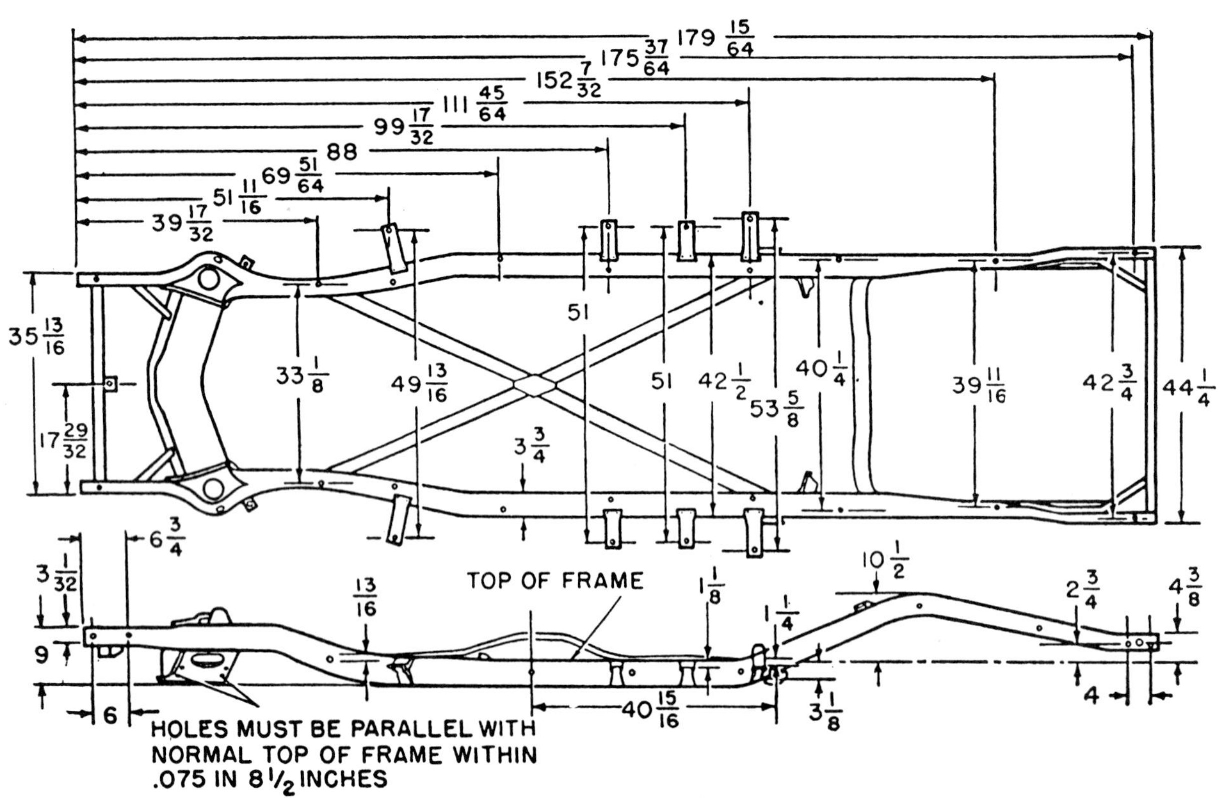

1952-1953 FORD STATION WAGON

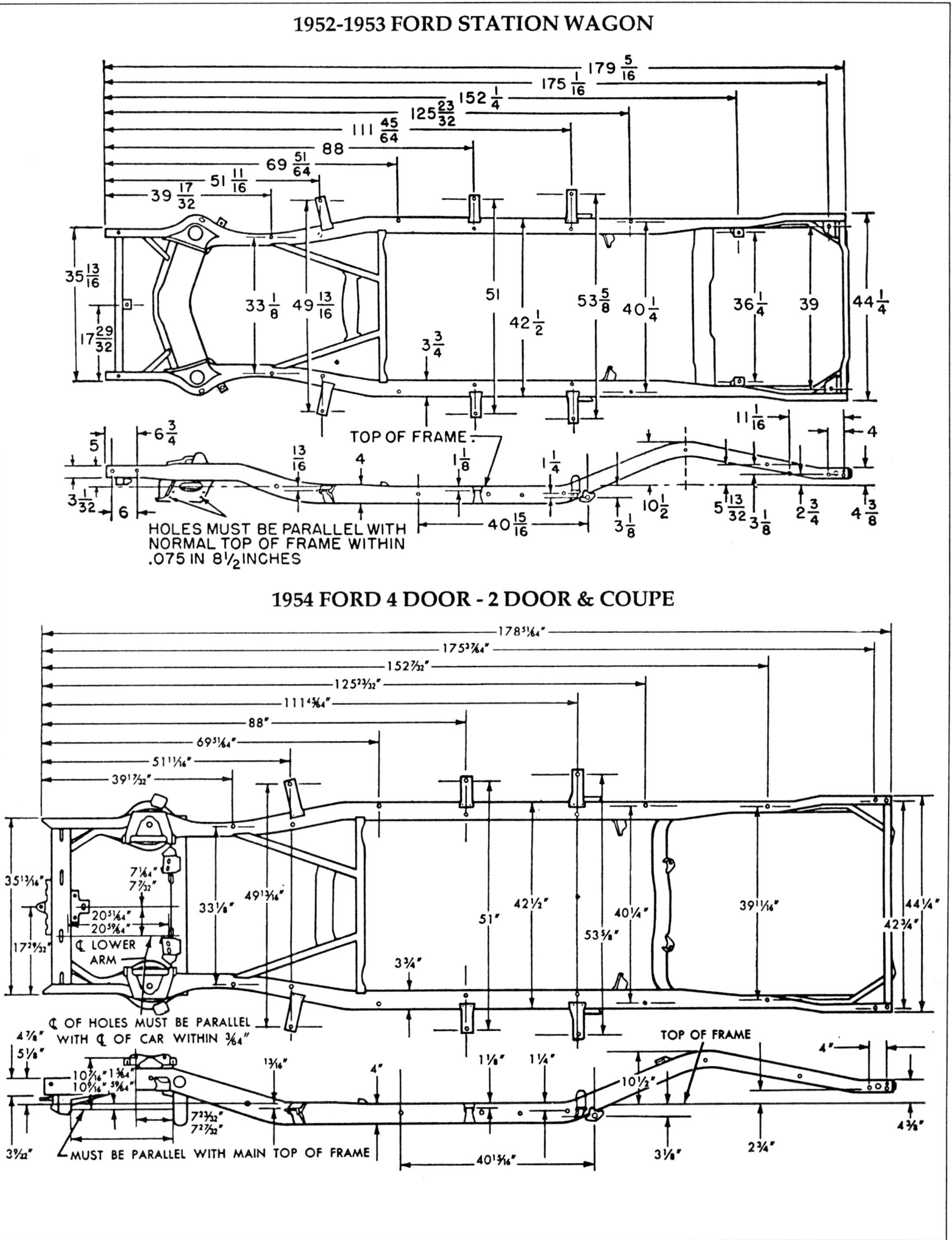

1949-1951 MERCURY (ALL)

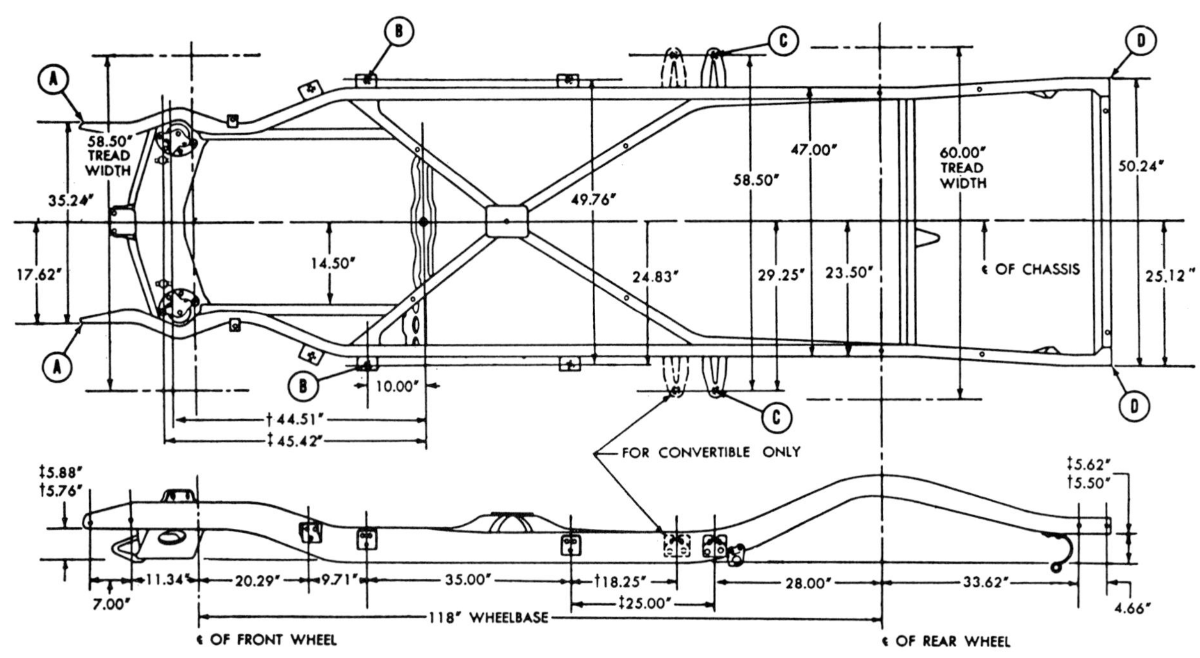

1952-1954 MERCURY STATION WAGON

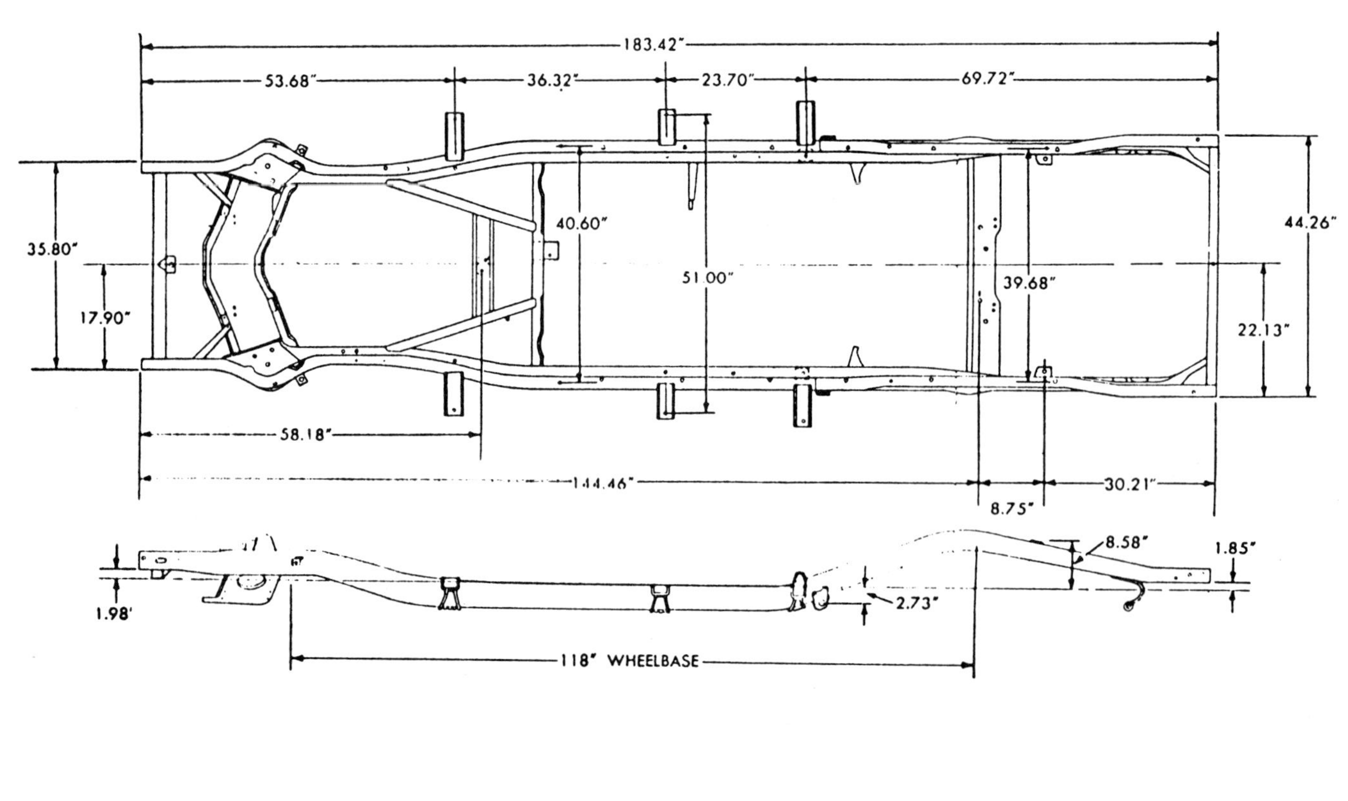

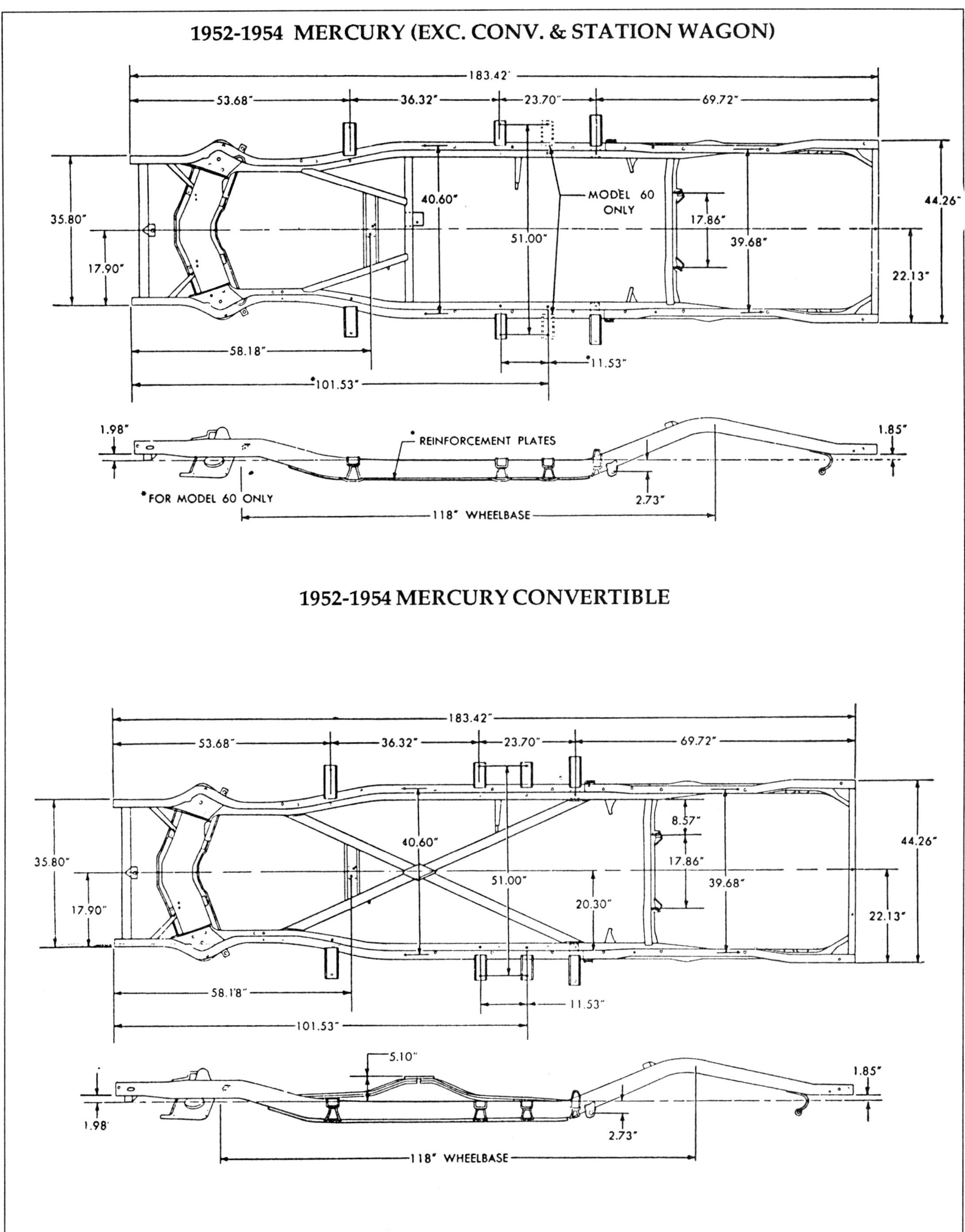
1952-1954 MERCURY (EXC. CONV. & STATION WAGON)
183.42"
53.68"
36.32"
23.70"
69.72"
MODEL 60 ONLY
40.60"
35.80"
17.90"
51.00"
17.86"
39.68"
44.26"
22.13"
58.18"
*11.53"
*101.53"
1.98"
* REINFORCEMENT PLATES
1.85"
*FOR MODEL 60 ONLY
2.73"
118" WHEELBASE
1952-1954 MERCURY CONVERTIBLE
183.42"
53.68"
36.32"
23.70"
69.72"
8.57"
40.60"
35.80"
17.90"
51.00"
20.30"
17.86"
39.68"
44.26"
22.13"
58.18"
11.53"
101.53"
5.10"
1.85"
1.98"
2.73"
118" WHEELBASE

CHASSIS SWAP

One of the most interesting ways to update the chassis of an older car is to simply swap the original body onto a late model chassis. Advantages to this include the automatic upgrading of brakes, steering and suspension components, resulting in a car with classic styling and modern ride/handling characteristics. This kind of thing got started a number of years ago, when Dick Dean started making full swaps in his southern California shop.

When doing a chassis swap on a Merc, Dean prefers to use mid-size Oldsmobile chassis of the late '60s to early '70s. These chassis have a 118" wheelbase (as do the 1969-'73 Pontiac Grand Prix), which is exactly what the 1949-later Mercury has. But Dean goes everyone else a step better. While it is possible to just drop the Mercury body onto the bare GM chassis, Dean cuts out the original Mercury flooring and much of the Mercury firewall. The GM floor is cut from the donor body, and the GM firewall is trimmed around the perimeter, leaving the steering and brake pedal assembly in place. In essence, only the wiring that is fastened to the GM fender splash apron is removed and laid over on the GM engine. Even the seats and gas tank of the GM unit are left in place.

The bottomless Mercury body is then lowered onto the GM floor and the Merc rear fenders aligned with the GM wheels. This places the firewall in almost exact original Mercury position, so the firewall is trimmed to allow the body to set down on the GM flooring. When everything is aligned, the Mercury body is welded to the GM floor, and the trimmed GM firewall is welded to the Mercury body. Instant swap. The frame horns must be modified to accept the Mercury bumper (or whatever bumper is to be used), and the radiator core support is welded to the crossmember. Mercury splash panels are trimmed to clear the GM suspension.

This frame swap can be accomplished very rapidly, and the results are gratifying, especially since the late GM seats can be used, as well as the entire steering assembly and brake assembly. Even the full wiring harness at the firewall can be modified to fit the early dash. With some careful trimming and fitting, the late GM heater/air conditioner will fit under the dash.

Obviously, the same kind of frame/flooring swap can be done on other cars. The key is to match wheelbases. This swap is especially appealing to someone with an early body that has a badly rusted floor. Such was the case with Paul Beckstrand's 1950 Mercury,

which had a very rough body in the flooring area. Badly rusted, it was a perfect candidate for a frame/floor swap.

The Mercury wheelbase is the same as a 1977 Oldsmobile 98 Regency four-door, and a Buick Park Avenue. Paul lifted the Merc body off the original frame, and hung the body in the air by means of a couple of come-alongs, suspended over the Oldsmobile chassis and floorpan. Working from the rear wheel location, he trimmed the Olds floor and Mercury floor sheetmetal a bit at a time, trying the fit until he had the body channeled 4 inches below the Olds flooring. Paul did all the work alone, and reports it wasn't difficult, but it did require a lot of patience. He noted that since the Olds tread width is slightly greater than the Mercury, the channel may cause slight wheel/tire interference. One solution to this is a positive offset wheel.

Here are the prime subjects involved in a typical chassis swap for a Mercury, using a late model GM chassis. The entire GM floorpan and lower area of the firewall are used.

Step one is to remove the original Mercury chassis and suspend the body from an overhead beam. This makes the project much easier, as the late model chassis will be slipped underneath for trial fit, trimming and welding.

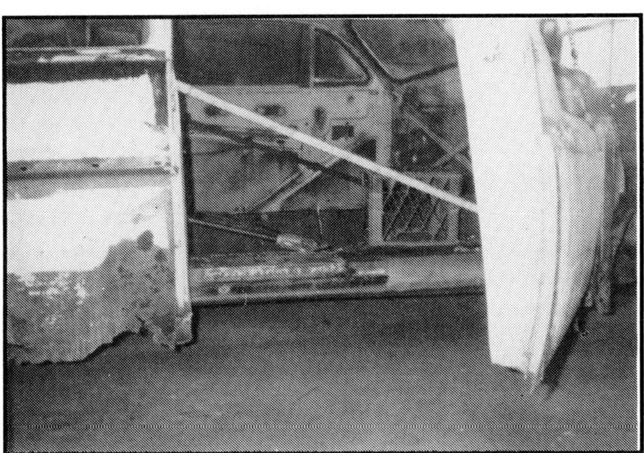

The chassis/floor swap is ideal for a body that has a badly rusted original floor. If there are plans to channel the body, now is the time to do it. This Merc is being dropped 4 inches over the frame and flooring, and new sections of sheetmetal added between the old body and the late frame. When doing this, the entire area around the perimeter of the body and floor must be welded solid.

Here, the quarter panel of a badly rusted Merc body has been cut away for later replacement. This exposes an area of body inner structure that can be beefed up and welded to the GM flooring. The idea is to make things stronger than necessary, and they will turn out just about right. After all welding is done, it is a good idea to fill all the seams and cracks with body caulk.

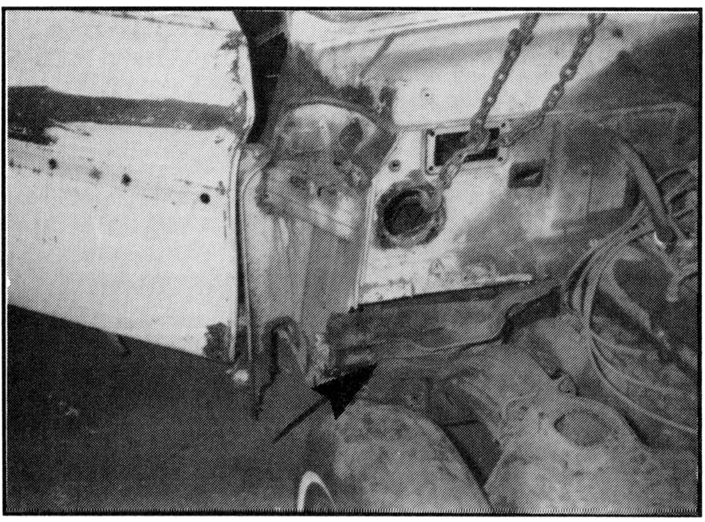

The body is suspended over the late GM chassis. Rear wheel opening is aligned with rear wheels. The chassis floor in this case is an Oldsmobile. The Olds firewall is trimmed just above the frame (arrow), and the Mercury firewall is trimmed to match. Then everything is welded together. In some cases, the entire GM firewall is retained.

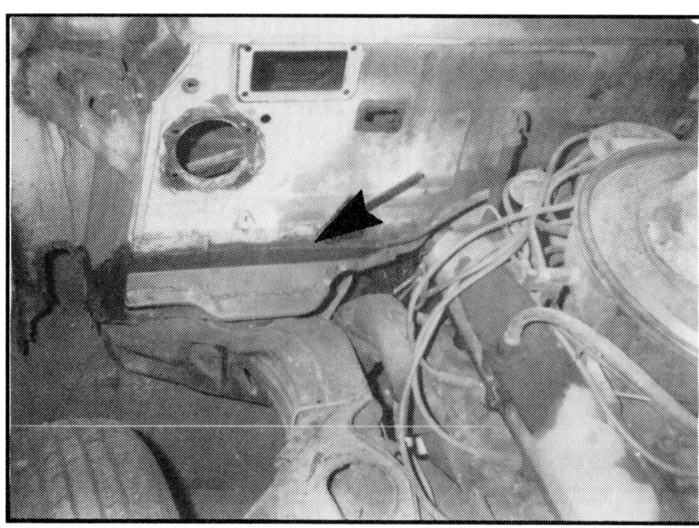

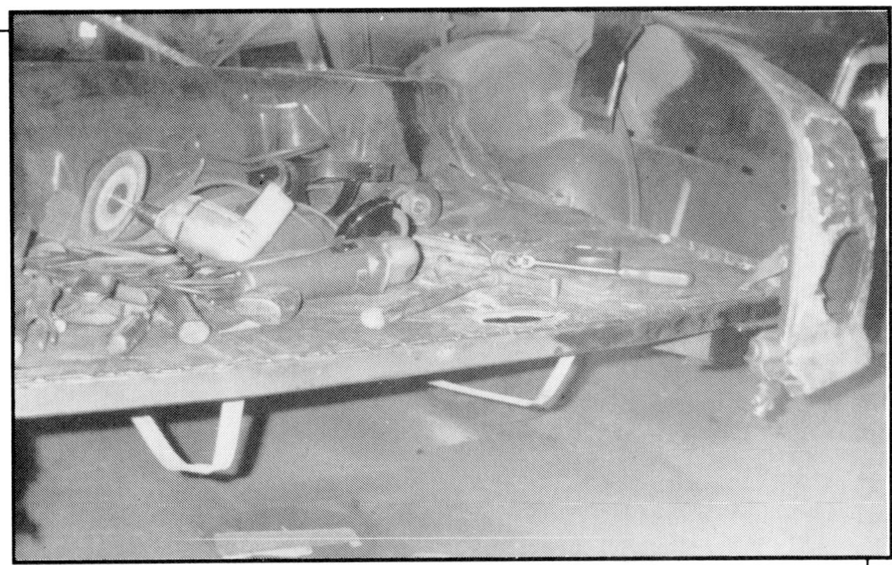

Note how the Mercury body rear lip is now well below the floor level, as a result of the channel job.

Sections of sheetmetal must be cut and fit between the Mercury body and the GM flooring. Again, it is vital that there be no gaps anywhere.

The inner wheel house is welded solid to the Oldsmobile floor pan, and small filler pieces of sheetmetal are added where necessary.

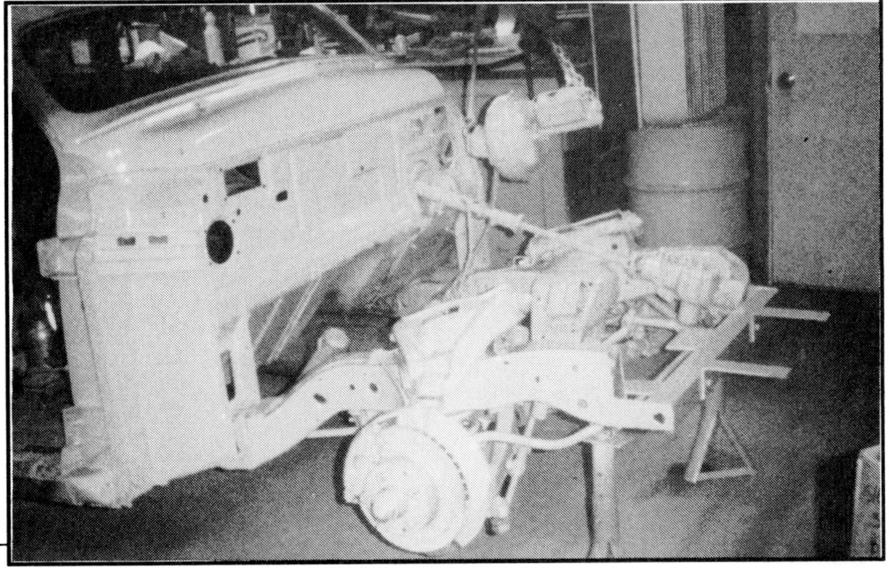

When it's all said and done, the Mercury is resting comfortably on an updated chassis. This gives a vintage car the advantages of late model disc brakes, upgraded suspension and steering components.

by Calvin Mauldin

SUBFRAMING

Slipping a late Camaro front under '49-'56 Fords and Mercs

Before we get into the "how-to" aspect of subframing these cars, let's address the situation. To subframe or not to subframe, that is the question. The majority of customizers know the reasons why subframing is neat, but if you are new to Kemps, or if you're sitting of the fence trying to make the decision, maybe this little review will give you the nudge you need.

If you want to stay traditional, '49-'56 Fords and Mercurys have serviceable front suspensions, and the flathead and Y-block can be worked on to provide ample horses. The braking systems are acceptable on these machines, as well. If this is the case, forget subframing.

But say you want to add power disc front brakes, a small block Chevy engine with turbo transmission, and to make your custom look real cool, cut the coil springs for the low look. Sure, the brakes can be done with machining and lots of headaches. A small block Chevy can be wrestled in with the help of an adaptor kit and lot of cutting, fabrication, and cussing. The cut spring coils are by far the easiest modification, but what you get as a result is a custom that rides like a wood wagon and gives you a bad case of double vision while cruising the nation's highways.

So, why not eliminate all of the above and use one of Flint's finest to solve the problems. Get all the goodies in one package: Disc brakes, a modern suspension, better choices of engines and transmissions, and the best part — lowering — while still keeping a smooth ride. Well, this is where subframing comes in. Got ya' convinced yet?

We caught Ken Ayres, owner of Precision Automotive, in Denison, Texas just as he was beginning a subframe job. Using James Isabell's chopped '54 Merc Sun Valley, and a '78 Camaro subframe as visual aids, this "how to" will guide you through the fundamentals. Before we start, remember the five basics: Think, measure, cut, think, measure!

Before any actual cutting is done, make reference measurements. Write it all down on something you can't loose, like the garage door, or the car's firewall.

Right-The original frame was cut just behind the idler arm. This gives lots of working room. It's also the point of no return, if you mess up.

Below-To arrive at what the final cut should be, there is one thing to consider. The '54 Merc frame is tapered, while the '78 Camaro frame rails are pretty much parallel (36-1/2 inches to 37 inches) from rail to rail. That's a problem.

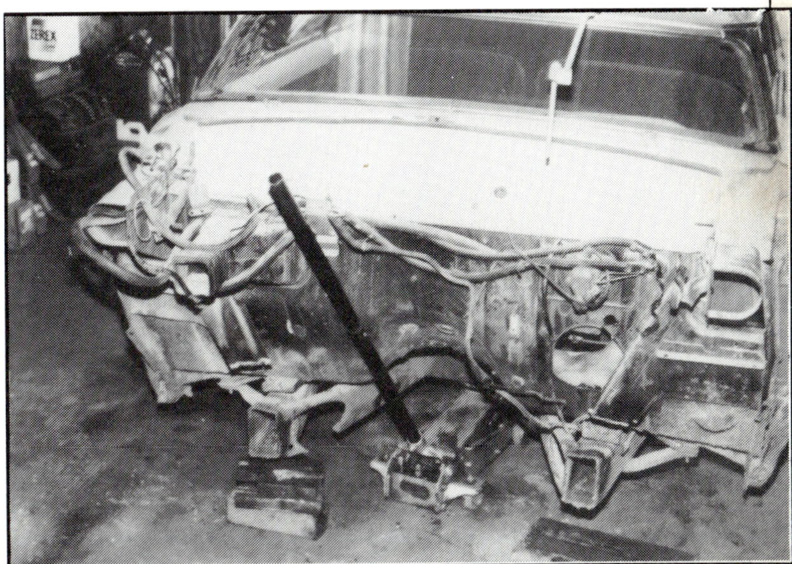

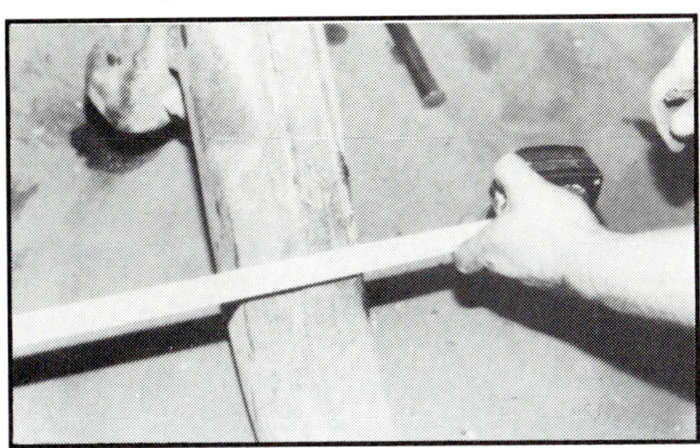

Right-The solution is measuring the Mercury frame from side to side until it tapers down to approximately 37 inches.

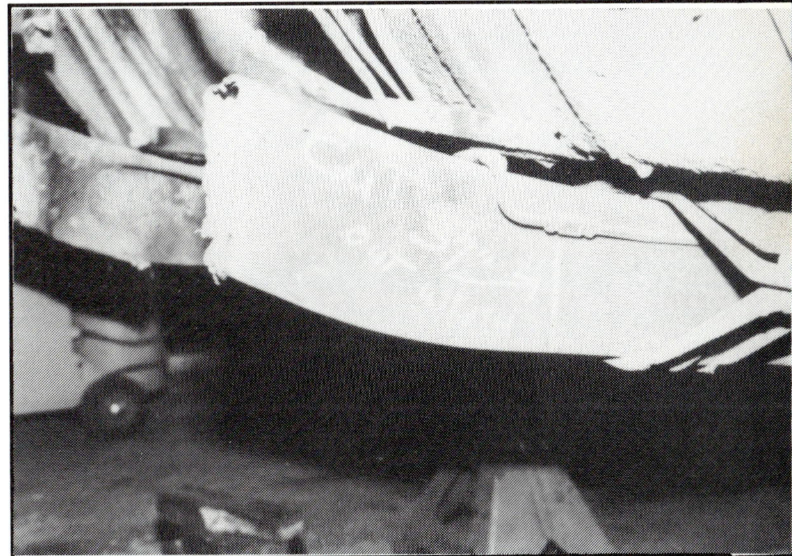

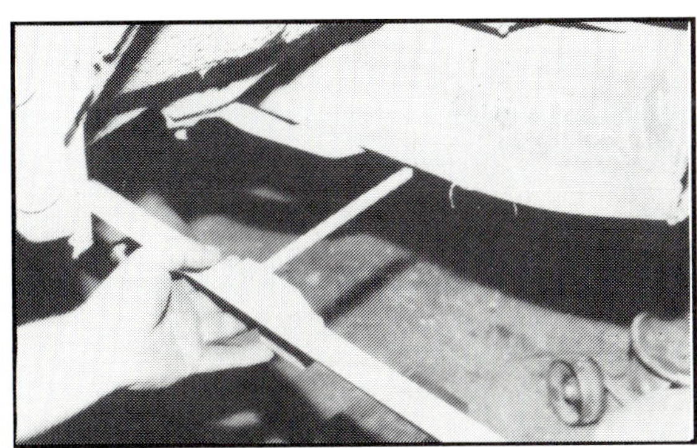

Above-Use a Trame gauge, because accuracy is a necessity in an operation such as this. Outside center hole in the rear axle is used as a reference point. The '54 Merc measures 93 inches from rear axle center to cut line, both sides. In other words, the original frame is square.

Right-This is the '78 Camaro front end awaiting a trial fit. On '49-'56 Fords and Mercurys, the 1970 and later front-steer unit is recommended. For the '48 and older, it's better to work with the rear-steer '67-'69 Camaro.

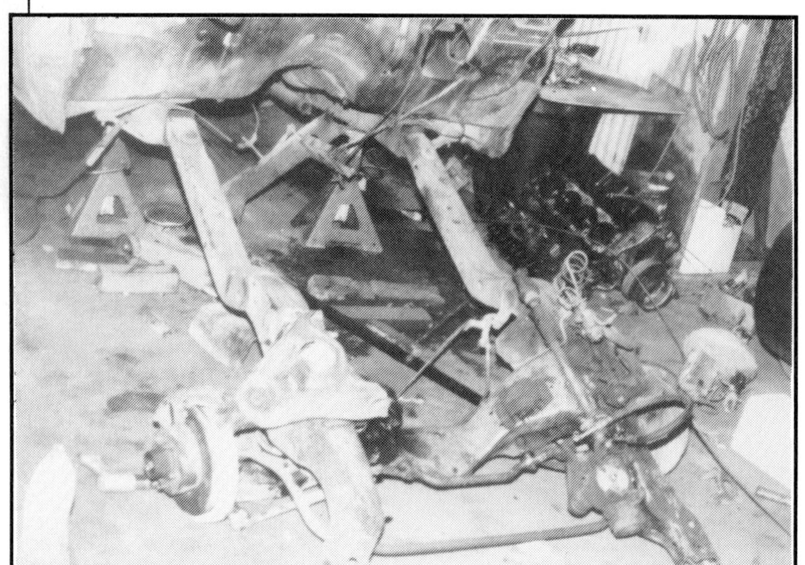

Left-This stage of progress makes you wonder if the thing will fit. Actually, this step helps to determine what has to be cut from where before proceeding further.

Below-The Camaro body mounts are trimmed away for clearance.

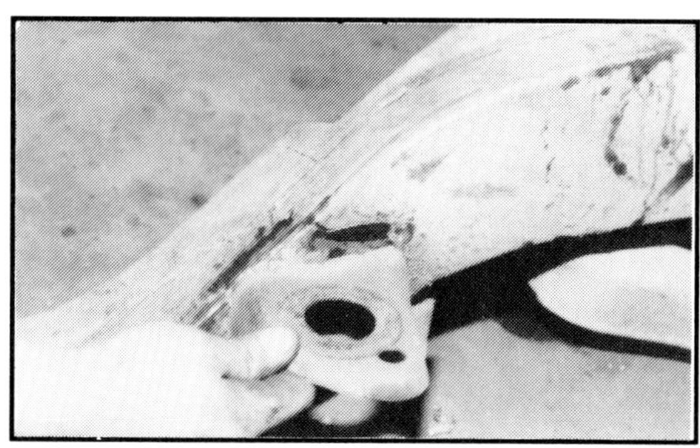

Left-A little over a foot of the rear section of the Camaro subframe was removed. This, of course, will differ on other brands of cars.

Above-Inboard sections of the Mercury frame rails were cut away. The slots allow the Camaro subframe to slide into the original frame.

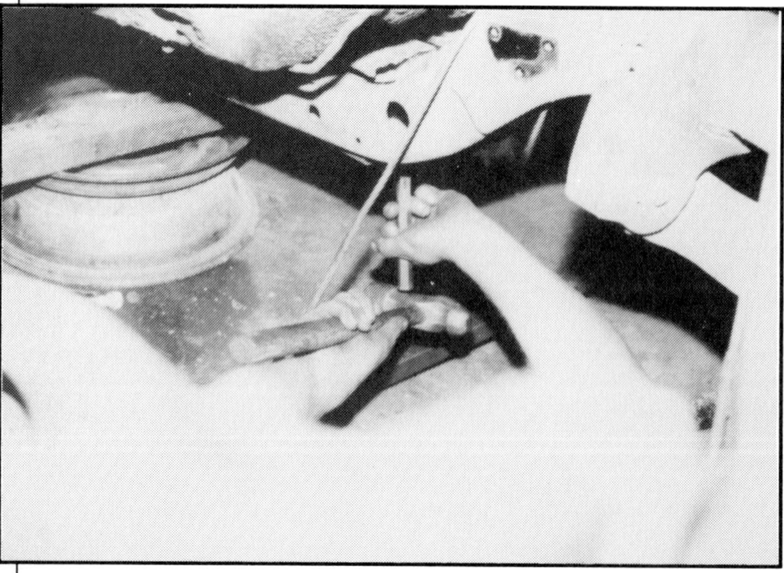

Left-With the subframe in place, a bit of blacksmithing is needed to get a snug fit. Note the slotted Mercury frame rails.

Right-Here Ken Ayres, owner of Precision Automotive, rechecks his handiwork. The two units have just been tacked together. The Trame gauge tells Ken the Merc still has its original wheelbase, both sides. This eliminates dog trot, or the embarrassment of looking like you're passing yourself going down the highway.

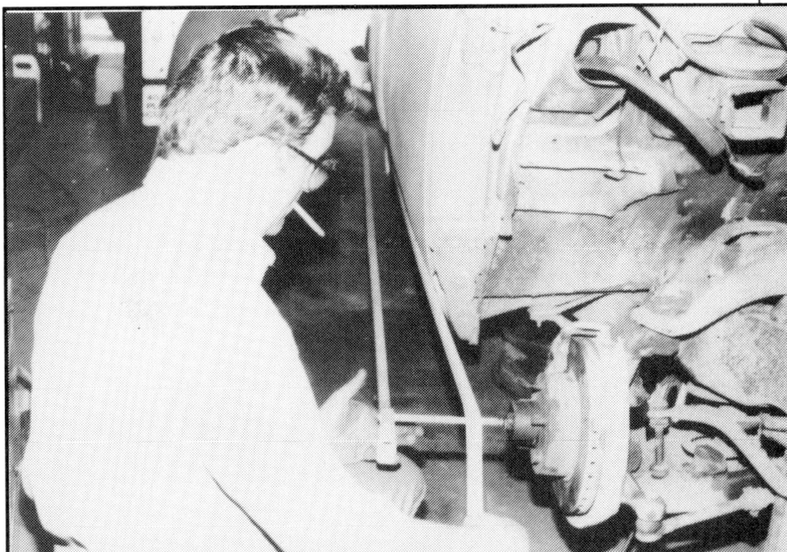

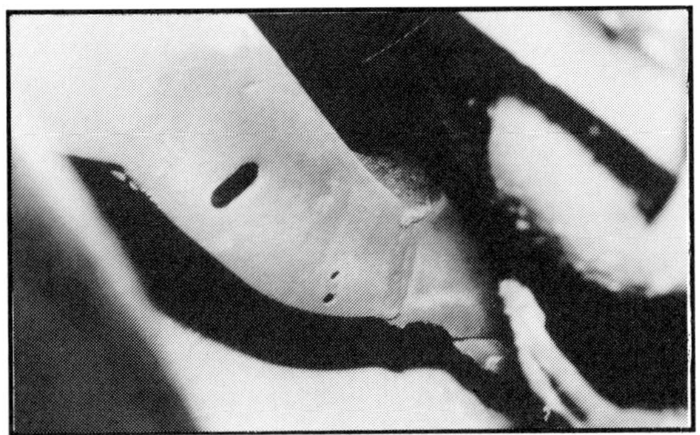

Above-The final welding has been completed, and with some grinding it will be hard to detect where the two frames were joined.

Right-You'd have to run into a speeding locomotive to make this frame move. Ken butted the subframe to the original for insurance.

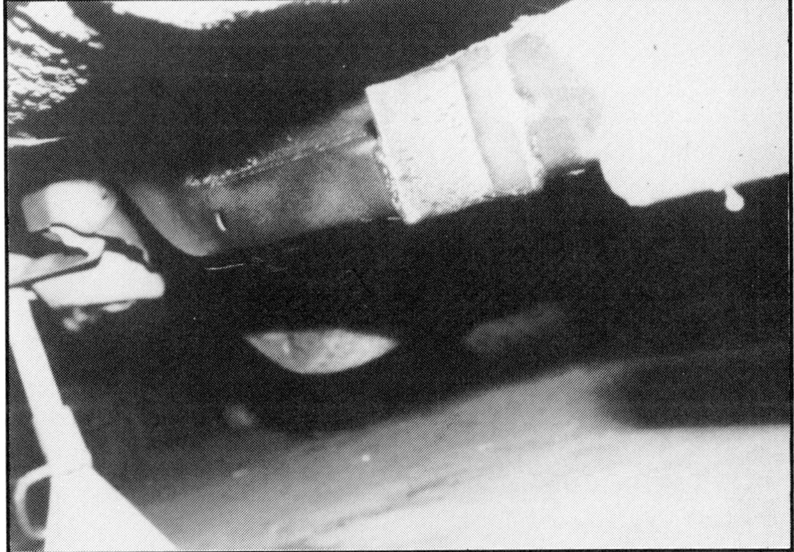

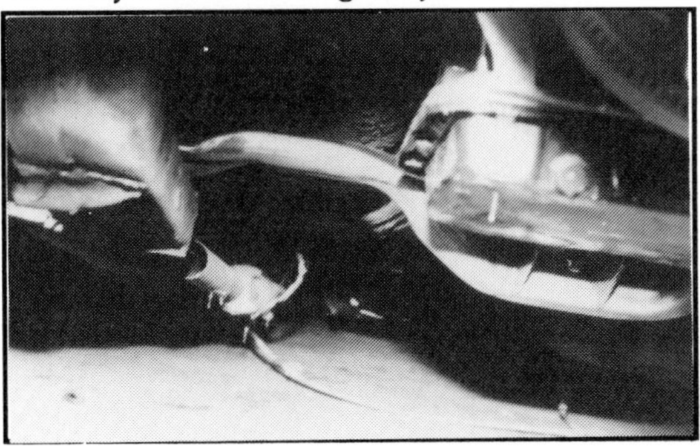

Above-The Camaro crossmember was retained. The more factory parts utilized, the less fabrication time required.

Right-The freshly subframed Sun Valley tests the shop floor for the first time. The wheelbase is measured once more. Ken keeps the replacement clip as close as possible to its original geometry.

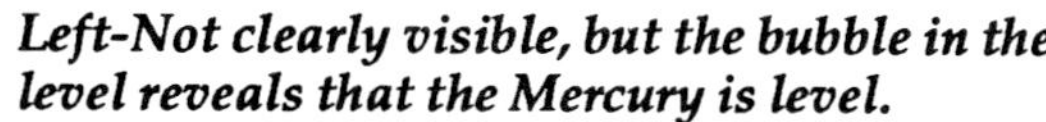

Left-Not clearly visible, but the bubble in the level reveals that the Mercury is level.

Below-Measuring from the floor to the front rocker panel shows a distance of 8 inches. With the installation of the small block Chevy, the body will come down an additional 3 inches.

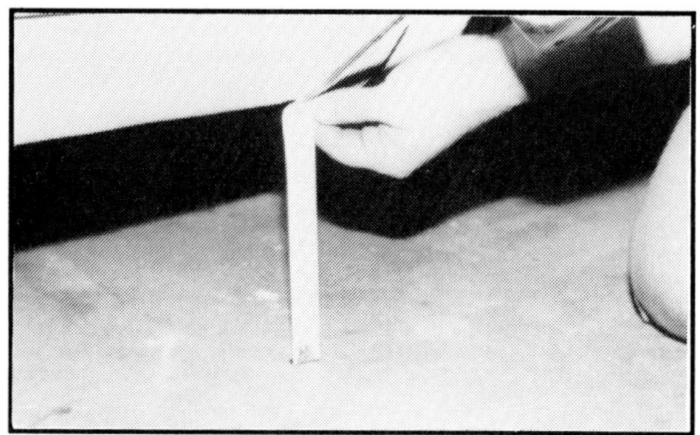

Left-To give ample clearance between the engine and firewall, the engine mounts were moved forward 3 inches. Is this mill dressed, or what?

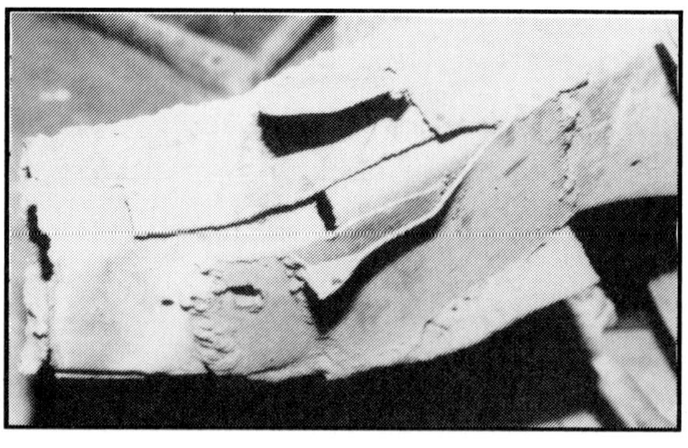

Above-This is "how not to subframe." Poor welding on this shadetree work was an accident looking for a place to happen. Doing it the right way will help prevent this kind of ugliness.

Left-After Ken finished his part of the project, the engine was plucked and the Mercury was trucked to David Guymon's Custom Emporium to be fitted with the front sheetmetal. Quite a bit of trimming was needed to get the inner fenders to fit over the Camaro frame. Note the relocated engine mounts.

Right-The biggest problem with subframed Kemps is what to do about support for the radiator, and if you use 'em, where do you hang the bumpers? David took care of this headache by attaching a '69 Ford C6 transmission cross-member. Neat idea, and very little fabrication.

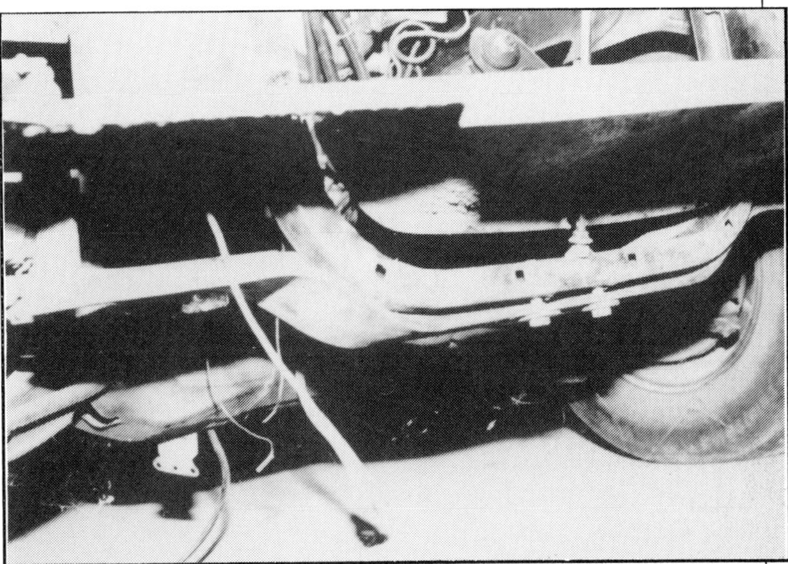

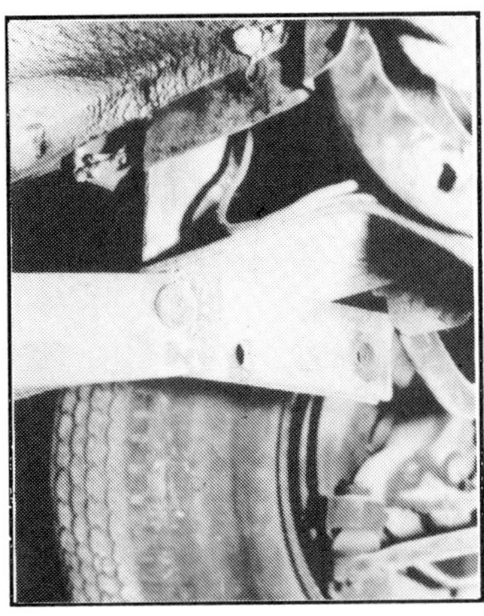

Above-With the bumper propped up, it is evident that a simple bracket will be all that's needed for mounting purposes.

Right-A pair of '52 DeSoto bumper brackets were cut, then angled downward.

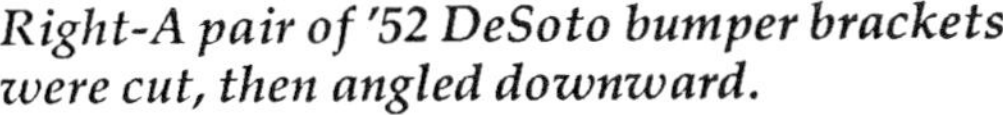

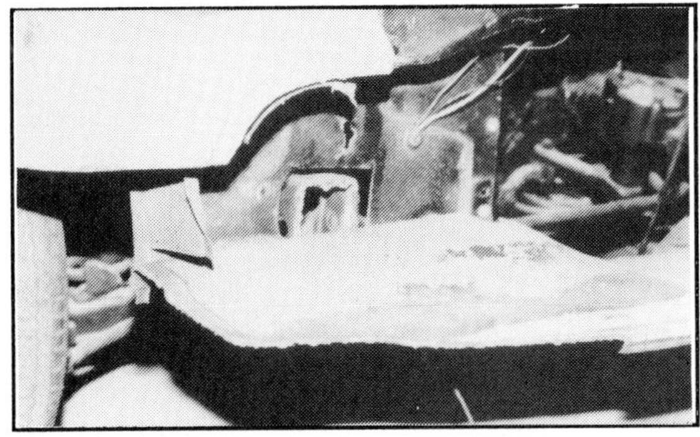

Above-Slots were cut to clear the front portion of the Camaro subframe. Lower splash panel is a trimmed '54 Ford item.

Right-Ready for some Guymon customizing. The '54 Sun Valley will have end gaps filled with the lower panels from '53 Chevy front fenders. Propped up for visual check is a '52 DeSoto front bumper.

by Rich Johnson

MAIL ORDER MERCS

Regardless of what happens in the future of custom cars, the '49-'51 Mercs are always going to reign as royalty. The problem is that on this planet there are only so many surviving Mercs to be salvaged for parts and pieces. And the number of original Mercs is obviously far below the number of folks who want to have one. That being the case, it was only natural that the aftermarket would respond with a load of goodies to be used in the building of these highly sought after models.

One of the aftermarket companies that has risen to the fore in the Merc parts department is P&J Automotive (6262 Riverside Dr., Danville, VA 24541; 804/822-2211). And the kind of stuff they have come up with will answer a lot of builders' prayers.

Take, for example, their '51 Merc chassis, available in Stage 1, Stage 2, and Stage 3 versions. For $2995, a builder can order up a Stage 1 chassis that features a 2x4 tubular frame with engine and transmission mounts, front crossmember with tubular upper and lower control arms, and rear spring kit, as well as all body mounting points.

The Stage 2 version includes all of the Stage 1 components plus a set of late-model GM spindles, rotors and calipers, steering gear assembly, front springs, and a late-model GM rearend assembly. Total price is $3995.

For the builder who wants even more, there is Stage 3. A price tag of $4495 will buy everything already discussed plus front and rear gas shocks, all brake lines run, a remote power booster, and a steering shaft kit. About all you need is a set of tires and

Have a look at this mail-order '51 Merc coupe body, made of fiberglass and featuring hinged and latched doors, hinged hood and deck lid, front fenders and rear skirts, as well as dashboard and garnish moldings. Coming soon are '49-'50 coupe bodies and '49-'51 convertibles.

wheels to make this a full rolling chassis, and everything is set to receive an engine and transmission.

Once the chassis is taken care of, a body is naturally the next item of concern. Well, in case there aren't any available steel bodies in your neighborhood, P&J Automotive has you covered with fiberglass versions. First to be introduced was the '51 Merc coupe body, made of fiberglass and including all the necessities. The body package costs $8950, and features doors that are already hinged and latched, a hinged deck lid, fender skirts, front fenders, a hinged hood, dashboard and garnish moldings.

Those are the basics, but there is a bunch of optional equipment that can also be ordered. The option list includes power windows that can be factory installed, power door releases and a power trunk lid release, already installed for reasonable prices. Other options include air conditioning/heater units, A/C condenser and compressor plus hose kit, a defroster kit, and electric cooling fan.

Also in the works are '49-'50 Merc coupe bodies, and convertible bodies of the '49-'51 variety. So, now it's possible to have a mail-order Merc delivered to your garage. Battery not included, some assembly required.

Three different levels of chassis can be bought from the P&J Automotive catalog. At the top of the list is a complete chassis that features late-model GM front components and rearend, gas shocks, all brake lines run, remote power booster, steering gear assembly and shaft kit, GM front spindles, rotors and calipers, etc. Installing a set of tires and wheels makes it a complete rolling chassis, then it's ready to receive an engine and transmission.

WINFIELD'S 'GLASS MERC

Gene Winfield is one of the pioneers in the custom car hobby, and has been responsible for some absolutely awesome projects in the past. Now he has come up with another winner, an all-fiberglass chopped Merc body, and it's a beaut! The steel-reinforced body comes with doors, hood, and decklid aligned, hinged and latched. A full-length fiberglass floor solves the problem of a rotted floorpan. Examples of the craftsmanship are the doors, which feature inner and outer panels that are bonded together while the outer skin is still in the mold, to prevent warpage.

While all this sounds great, there's more. Gene will sell the whole Merc (including skirts) as a complete package. Or he'll sell individual components separately, for those who may not need an entire Merc but do need specific parts.

All this is available from Gene Winfield Rod & Custom Construction, Inc., 7256 Eton St., Canoga Park, CA 91303; (818) 883-2611.

by Jim Genty

LOWERING FRONT SUSPENSION

Remember the "good ol' days?" That's all folks talk about today. Remember lowering your custom by heating or cutting the coils, then sneaking up on every railroad track and pothole in town? Ouch! And forget about leaving town. The longest trip we took back then was to the drive-in theater on the outskirts of town.

Well, the folks at J & M Enterprises remember those days, having dodged their share of potholes in the '50s. They have come up with a better mousetrap that will carry your cool '50s ride into the '90s. The JAMCO Suspension Package is an economical alternative to heating and cutting coils, and a perfect companion to dropped spindles, if you really want to slam it to the ground.

We went to CJ's Automotive in Red Bud, Illinois to document the installation of one of the deluxe kits on a 1950 Ford club coupe. CJ was using the F3 kit, which consists of a new computer-wound coil spring with a built-in 3-inch drop and a 30% stiffer spring rate, which helps offset body roll that '50s Fords were so famous for. In addition to the coils, the kit comes with brand new shorter A-arm bumpers (don't cut the old one — it defeats the purpose), new sway bar bushings, and two new gas shocks. Let's follow along with the procedure.

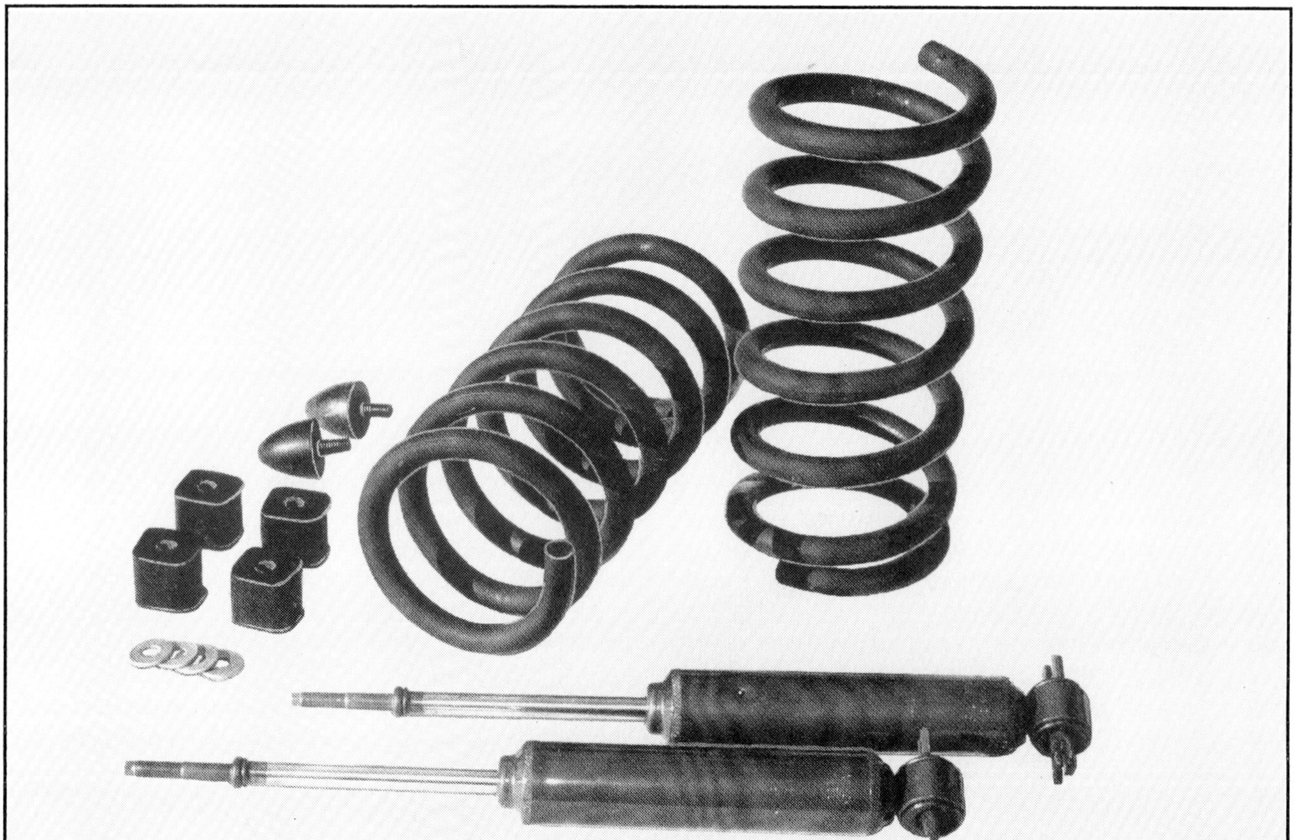

This is the starting point. Measurement from bottom/center of the fender lip is 28-3/4 inches to the ground.

CJ's first order of business was to get the front end jacked up in the air and secure the car by placing jackstands under both sides of the frame just behind the front tires.

Close up shot of A-arm area shows the 40-year-old suspension products in their proper locations. Coil springs, rubber bumper, shocks and sway bar with rubber bushings.

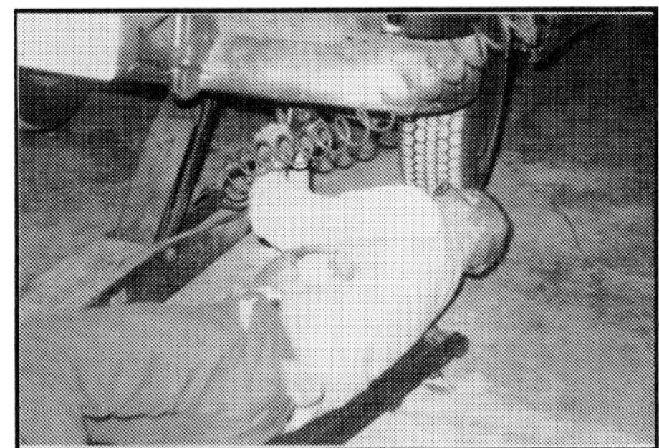

CJ attacks the bolts with modern air tools, but you can accomplish the same job with simple hand tools and a little muscle.

After years of being wedged into the A-arm channel, a pry bar may be necessary to remove a stubborn sway bar.

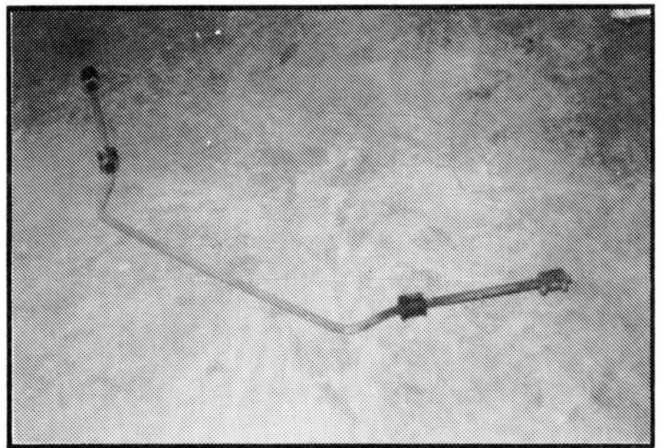

Sway bar removed and showing age-hardened rubber bushing. Years of road grime and surface rust will have to be removed prior to attempting installation of new bushings.

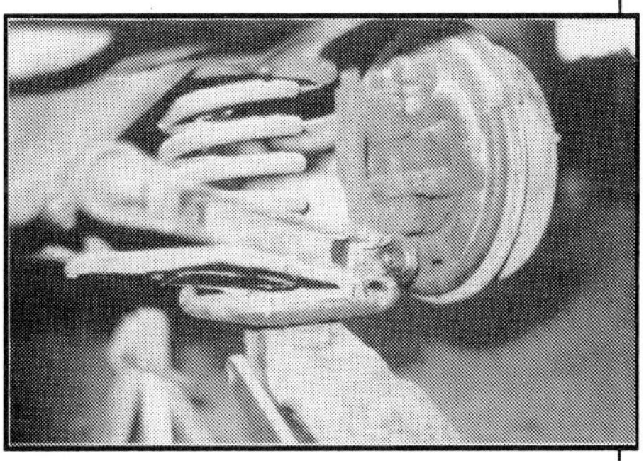

Remove shock absorbers and place floor jack under the A-arm to unload the downward stress, as we get ready to remove the lower spindle support bolt. Suspension is jacked snugly against frame at this point.

Left-Once the spindle bolt is removed, slowly lower the jack until coil spring is free and ready to be removed. Note: There is no tension left in the coil at this point.

Right-This photo shows the differences in the old vs. new rubber bumper. Do not cut the old ones, you need the cone shape in order for the bumper to work properly.

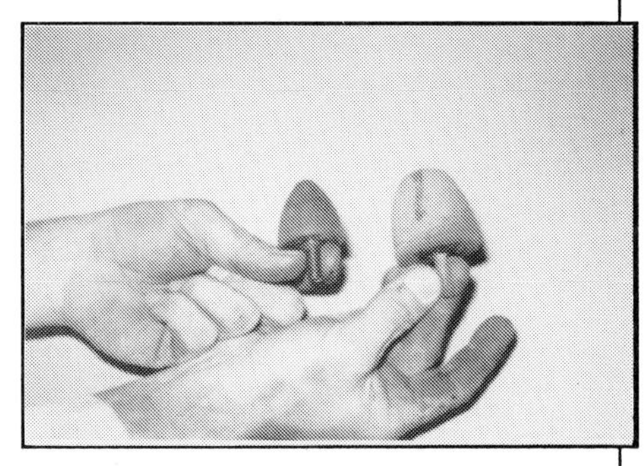

The sway bar was taken to the wire wheel and thoroughly cleaned up. Safety glasses and gloves and a heavy shirt are a must for this operation.

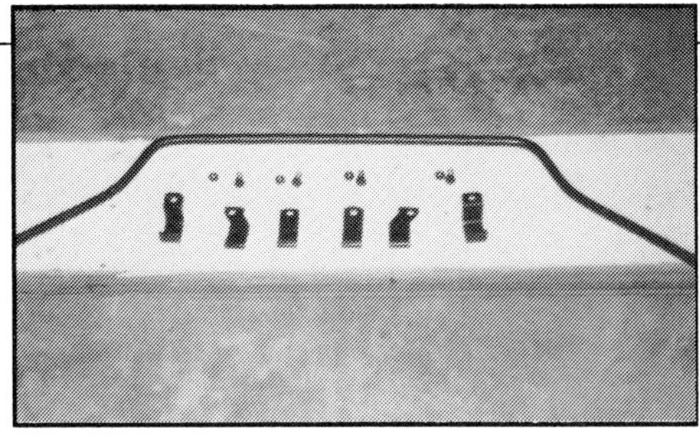

The sway bar and various clamps were then painted with a gloss black epoxy paint for a smooth, clean look.

This photo shows the difference in height between the old coil on the left and the new coil on the right. Note the difference in windings. This is more than just a shorter coil.

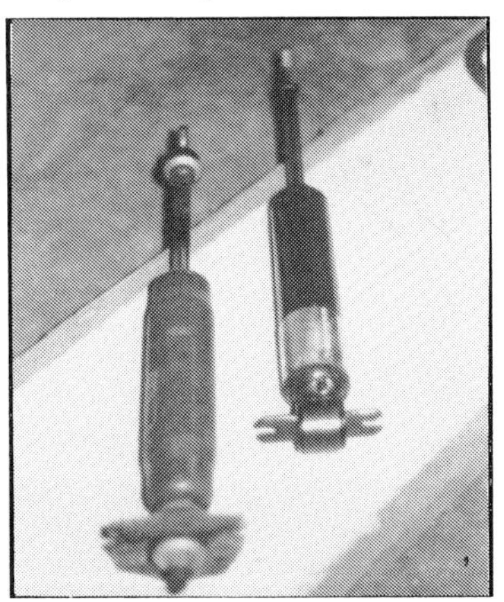

This photo shows the old shock on the left with the factory shock clamp. Shock on the right is the new gas shock with the current type mount. The original bolts are used with a shoulder washer. Do not attempt to use the old mounting bracket, it will not work.

CJ bolts the new shocks in, using the shoulder washers. Note: The shocks must be installed with the A-arms in a loaded position.

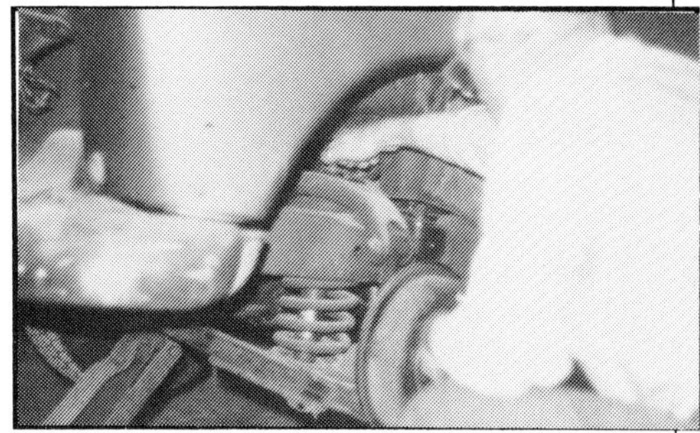

Here, CJ finishes attaching the spindle support bolt and snugs down the top of the shock.

The tire is now back on the ground. Bottom of fender lip-to-ground measurement is now 25-1/2 inches, with plenty of suspension movement. CJ runs a 14-inch wheel with a low profile tire that does not fill the fender opening the way a 15-inch wide white wall tire would.

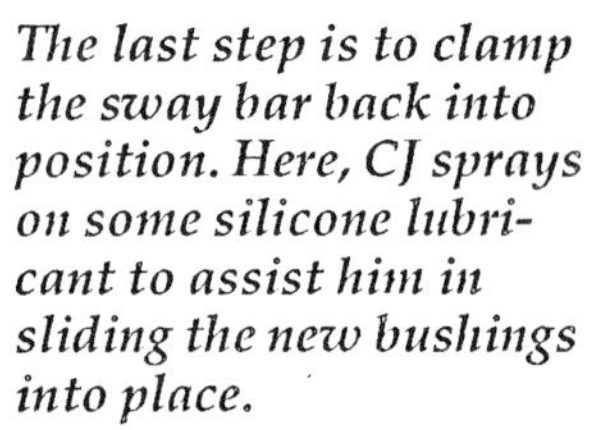

The last step is to clamp the sway bar back into position. Here, CJ sprays on some silicone lubricant to assist him in sliding the new bushings into place.

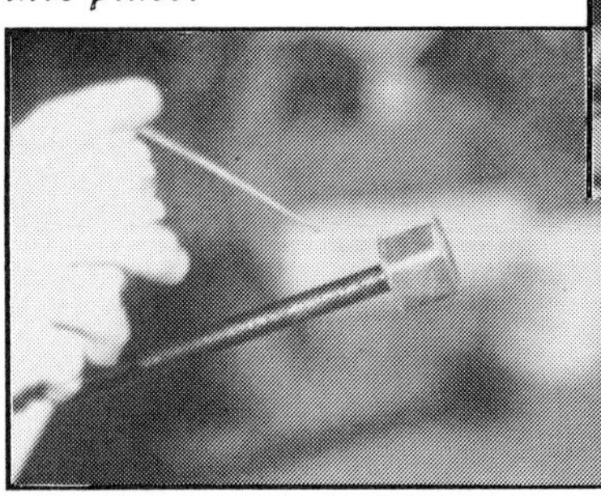

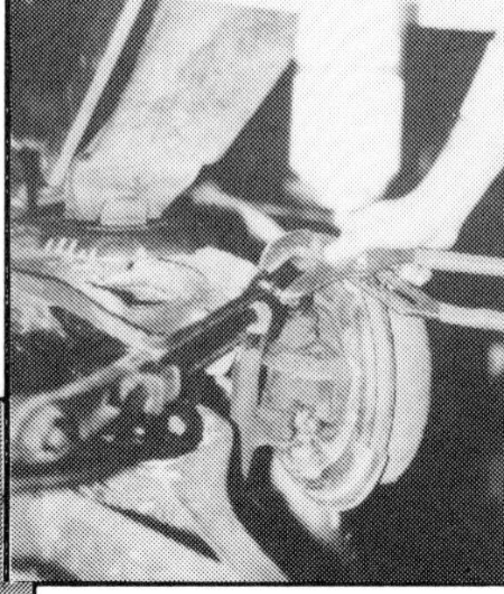

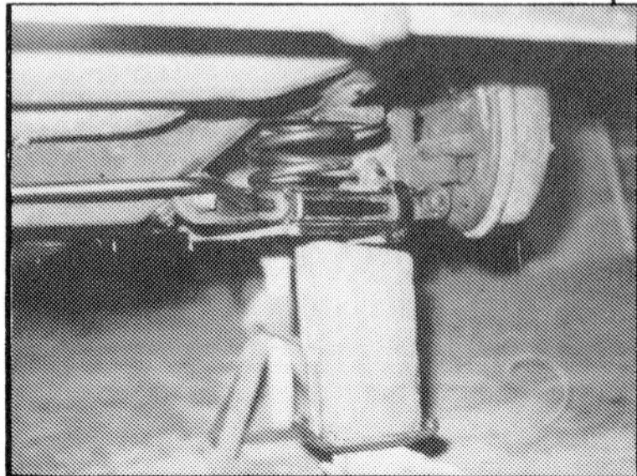

SOURCES:
J & M Enterprises
PO Box 510023
St. Louis, MO 63151
(314) 846-7057

CJ Automotive
1433 E. Market St.
Red Bud, IL 62278
(618) 282-6181

by Jim Genty

LOWERING REAR SUSPENSION

Back in the good ol' days, the first thing you got for your new wheels was a set of lowering blocks to get the back end of the car down. The ride quality was terrible — teeth jarring, kidney shaking, but you had to look cool, right?

Well, today folks still want to look cool, but they also would like some ride quality and stability to go along with those looks. That's where J & M Enterprises and their JAMCO Suspension Package come in. They grew up in the Fifties and remember that "old ride" very well.

To complement their computer-designed front coil spring package, they have come up with a similar package for the rear springs that, in stock form, delivers a 30% stiffer spring rate to reduce body roll when cornering. This package was then modified to give a 1", 2" or 3" drop, but still provide a good, secure ride quality.

Now, if you never go any farther than the local cruise spot, you might not feel the need to install new springs. But if you travel to any runs more than 100 miles away, you will really appreciate this ride.

In the photos that follow, a 1950 Mercury coupe, owned by Mel and Wanda Bauer of Hazelwood, Missouri gets the JAMCO rear spring treatment. Mel's car had been previously lowered in the rear in the conventional manner — 2" lowering blocks coupled with 40-year-old springs. It gave him the "look" but no ride quality to speak of. Mel decided to go with the JAMCO M-3 kit, consisting of new springs with a 3" drop, 30% stiffer spring rate, and new gas shocks. A cool look, with a Nineties kind of ride.

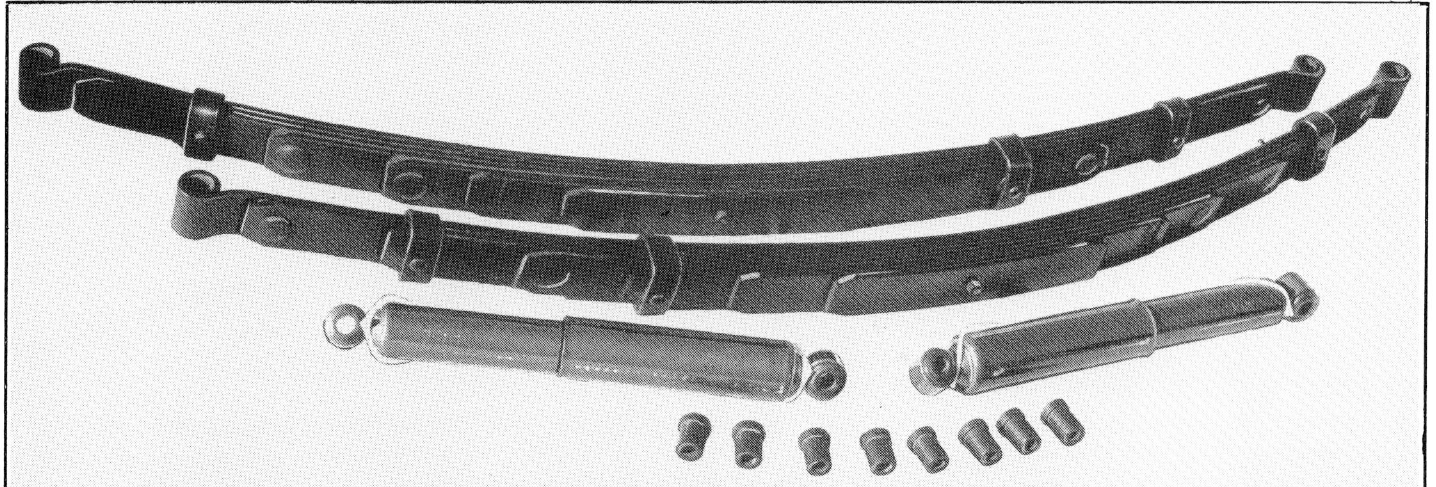

This is our subject — Mel's '50 Merc with 40-year-old springs, 2" lowering blocks, and no ride quality.

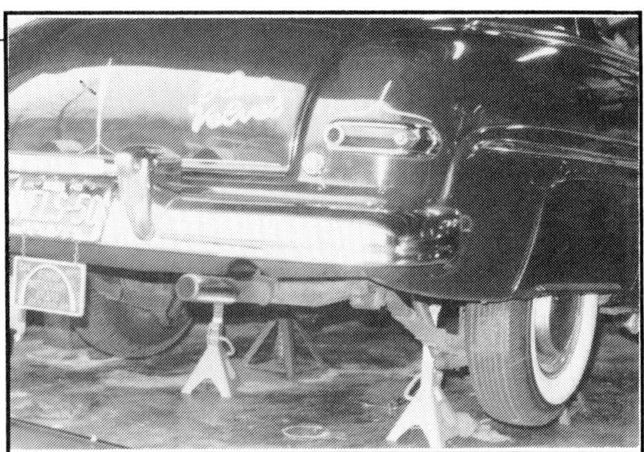

The first order of business is to get the car up on jack stands. Note that tall stands have been placed under the frame, and shorter stands under the rear axle housing. Don't forget to place a wheel chock in front and behind the front tire to prevent the car from moving.

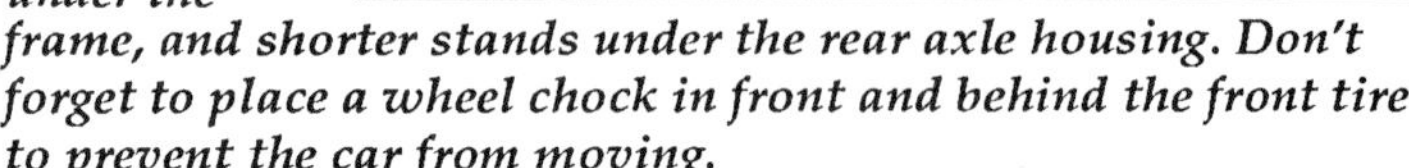

To give more accessibility and room in the working area, the fender skirts and the tire should be removed.

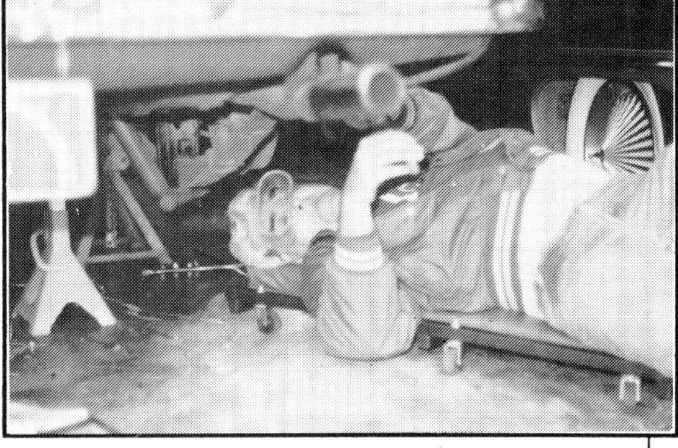

At this point, Mel crawled under the car with wrenches in hand and started loosening the old nuts. Safety glasses are a MUST in this position.

Take the shock loose first. This allows the rearend to drop down on the second set of jack stands when the springs are loosened. It is recommended to use a spray can of Liquid Wrench to help loosen the old nuts. Spray liberally, and give it about an hour to penetrate.

Take the rear mount loose first. Remove the nuts from shackles and pull down on the rear of the spring to unload the shackle. If you experience a lot of tension, try a pry bar and/or lower the jack stands under the axle.

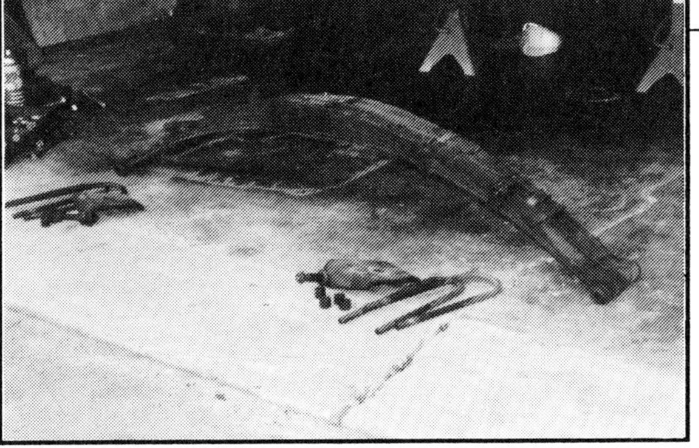

Once the front bolt is removed, the front end of the spring simply pulls out. A second set of hands helps. Here is the fruit of our labor — two very worn out springs, U-bolts, and shock mounting pads.

Note the broken leaf. We found numerous breaks as we checked out the old springs.

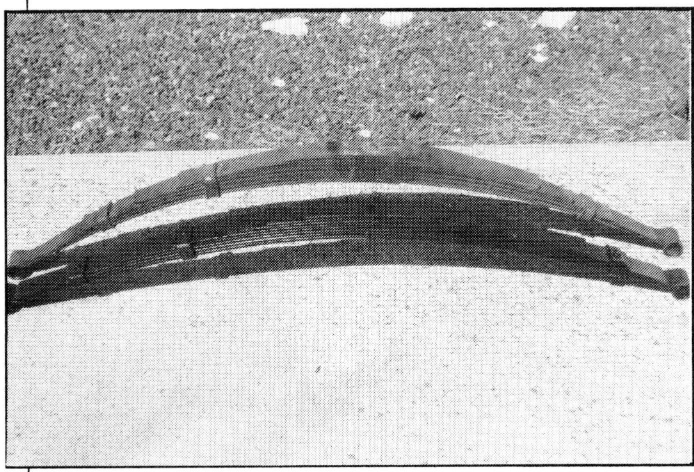

Visual comparison of old vs. new. The old spring is in the rear, and the new one in the front. Note the difference in arch. Lower, but better.

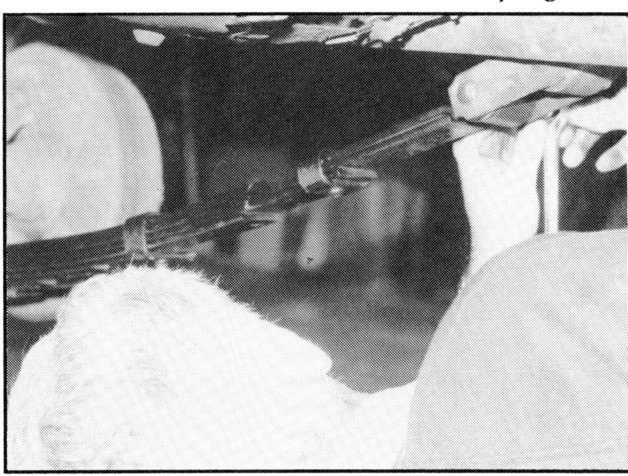

Reverse the mounting process by installing the front end of the spring first, then the back. A little WD-40 or silicon spray goes a long way here.

Note the visual comparison without the old 2" lowering block installed. U-bolts should be cut off or replaced with shorter units for clearance.

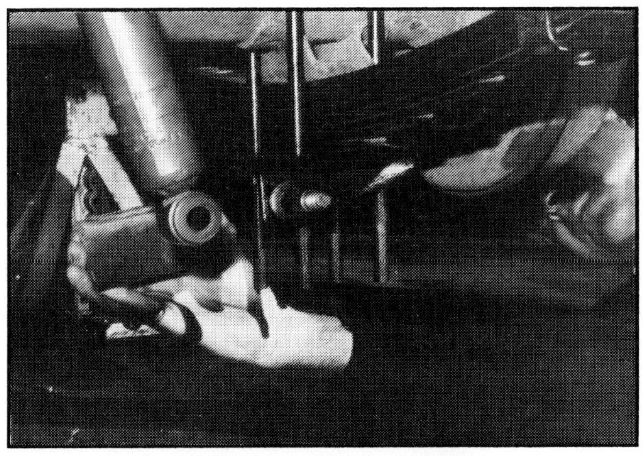

Install shocks and jack the car up so the rear shackle is forced back up into place. Replace tire, and remove the jack stands.

This is the new stance. A 3" drop without the use of lowering blocks. A 6" drop can be easily obtained with a combination of M-3 springs and 3" lowering blocks, and still have a great ride quality.

by Larry Kramer

HOW TO MAKE YOUR OWN LOWERING BLOCKS

One technique for getting the car down in the weeds is to install a set of lowering blocks. These are used on cars that have the leaf springs slung under the axle. The blocks are installed between the axle and the spring pack, effectively dropping the springs. Because the springs are attached to the frame or the bottom of the car, the whole chassis or body is lowered.

Use long blocks, and you get a lot of lowering. Use short blocks and you get a little. Take careful measurements of the car's clearances before deciding how long a block to use. If you bring the car down a lot, you might have to do some modification to the fenderwells to keep the tires from rubbing.

The following series of photos illustrates how to make and install your own set of lowering blocks. In this case, the blocks were being made for a GM rearend.

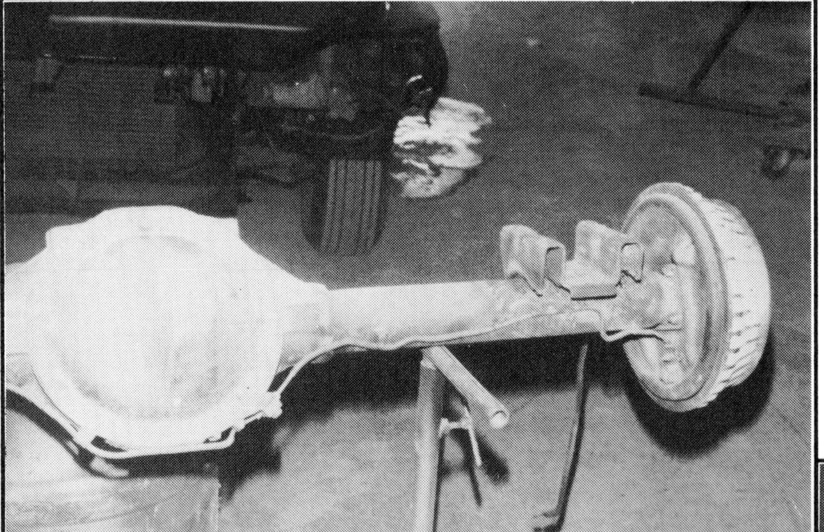

With the axle supported on stands, trim off all old spring mounting pads by using a torch. Take your time and be careful not to damage the axle housing. Leave all the metal on the housing when trimming the brackets away.

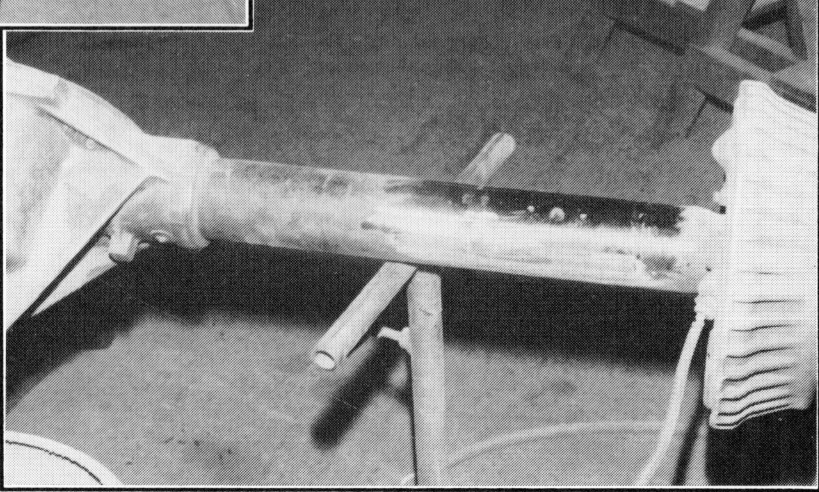

Using a grinder, clean the housing and make it ready for the new lowering block spring perches to be welded on.

Make the lowering blocks out of 2x4-inch rectangular tubing. The height from the radius to the top of the block (A to B) will determine your total drop. In our case, it is 5 inches. Start with a single length of tubing and use a hole saw the diameter of the axle housing to cut a hole through both walls at the center of the length of tubing. Then cut across the center of the circle left by the hole saw to separate the two blocks. This makes both radius cuts at once. Make top plates out of 2x4x1/4-inch flatstock.

Carefully position the plates on the end of each block and weld in place. Start with a tack weld to make sure the plate doesn't shift out of position, and then weld all the way around.

Use diagonal lines drawn from the corners of the end plates to locate the exact center. Drill the center hole for the spring alignment pin. Make the hole the same diameter as the pin to get a tight fit when the springs are mated with the blocks and bolted down with new U-bolts. A local spring shop will make a set of long U-bolts that are necessary after the blocks have been installed.

Before welding the blocks to the axle housing, grind down all the welds to make them smooth. Be careful not to grind all the way through the weld, or you'll have a weak bond between the block and the end plate.

Bottom left-Measure backing plate to block on each side to make sure it is the same on both sides. Then make minor adjustments to keep the distance between the holes in the end plate on the blocks equal to the distance between the locating pins on the springs.

Bottom right-Make the block parallel with the pinion. Square the back of the block and square the pinion flange. Double check all measurements and settings and then tack weld blocks to housing. To prevent heat-related warpage to the housing, weld in 1-inch intervals and then let it cool before welding another inch.

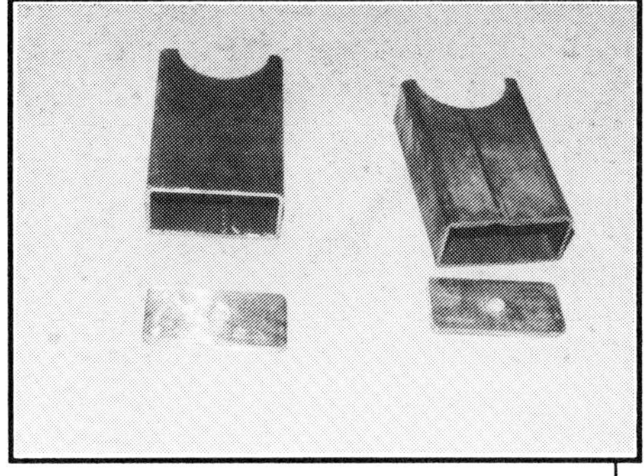

Drill the center holes in the 2x4x1/4 inch plates. Before drilling, use a center punch to help stabilize the drill bit so it doesn't wander all over the plate when starting the hole. A small pilot hole also helps eliminate the wandering bit problem.

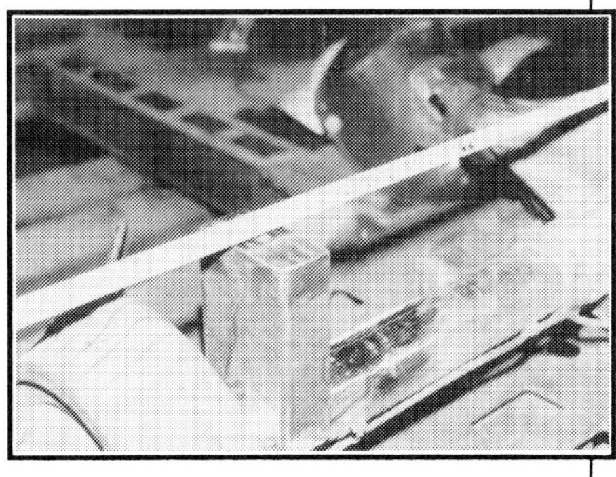

Measure the distance between the spring alignment pins, then match this measurement between the holes in the lowering blocks as they are resting on the axle housing. You want to make sure the distance between the centering pins of the springs is the same as that for the holes in the block end plates. It doesn't hurt to double check all measurements before welding.

by Rich Johnson

HYDRAULICS

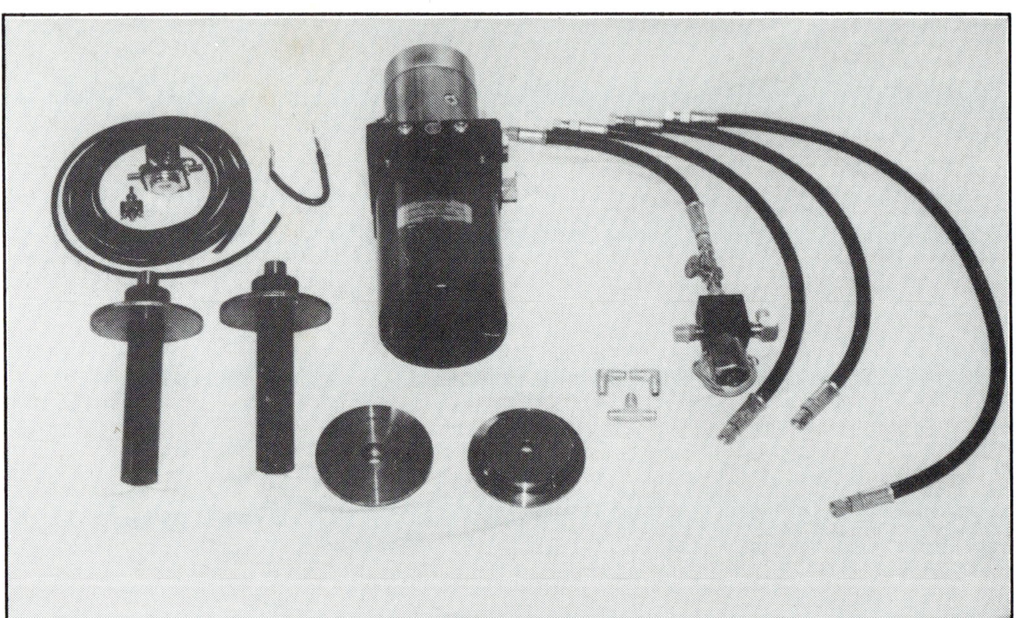

A paradox exists with custom cars. While they are neat to drive, being generally more comfortable than a hot rod, they look best when the car is real low. But a car that is down in the weeds sometimes has trouble motoring along under highway conditions, especially when rough roads or speed bumps come into play. So the question comes up, how do you have a car that can be safely and comfortably driven on the highway, but lowered for show? One answer is hydraulics.

Hydraulic suspension systems aren't new to the custom car market, but they are getting better and easier to work with all the time. Some very interesting concepts are being tried these days, but there have been a bunch of cobbled jobs come out of backyards in the past. Part of the problem was the fact that folks were left to scrounging parts and pieces of old aircraft or truck lift-gate hydraulic systems, and fabricating in cut-and-paste fashion rather than having components that were designed for use on a car. Not that we're opposed to the wrecking yard approach to hot rodding and customizing, but that doesn't always result in the cleanest installation and best engineered system. Now, however, it's possible to purchase ready-made hydraulic suspension systems directly from such places as Balls Hot Rod Parts (R.R. 4 Box 62-C, Syracuse, Indiana 46567; 219/457-2880).

Along with their components, Balls offers instructions for installation of a hydraulic system, as well as a list of Do's and Don'ts. Their schematic illustration has been included here to make it easier for you to visualize how the system goes together. It is also easier if you know what the components look like and how they relate to each other.

INSTALLATION

Use Teflon tape on all pipe thread fittings. Put fittings onto the dump valve and the pump. Make certain that all fittings are tight so they don't leak. Tighten fittings on the dump valve, then attach the dump valve to the pump.

Place Teflon tape on pipe thread sides of hydraulic cylinder fittings and tighten into cylinders.

Remove shock absorbers. Note: For back installation, leave shock absorbers attached.

Remove coil springs. Cutting with a torch will make it easier to remove coil springs.

Using a torch, cut a hole 1-1/2" in diameter in the center of where the shock absorbers were bolted.

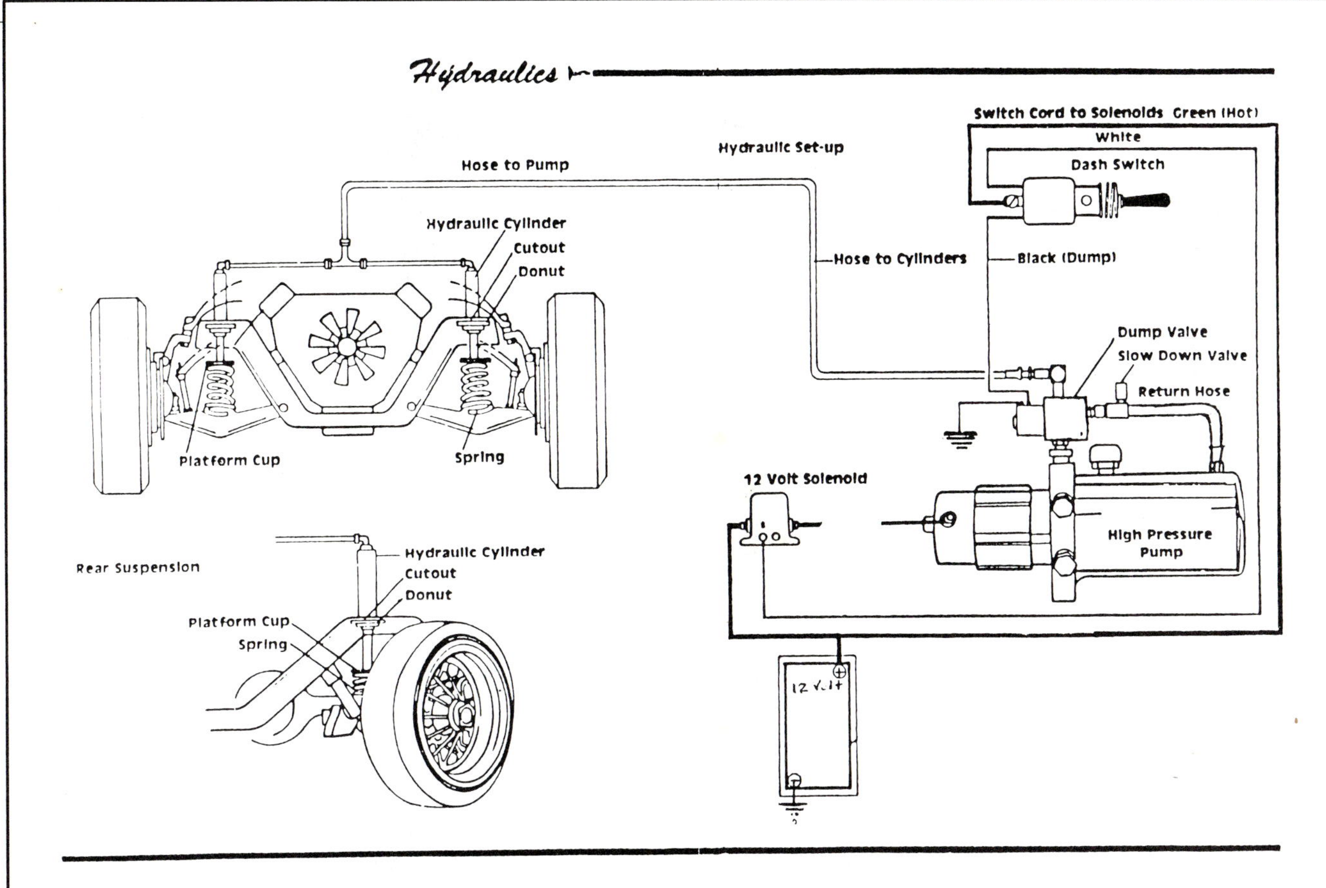

Cut coil springs. If the car is to be raised, leave 5 turns on the coil spring. If the car is to be lowered to ground level, leave 3-1/2 turns on the coil springs. Make certain the flat side of the coil spring is facing upward, and the cut side is facing down.

Slide the donut (large washer) over the hydraulic cylinder and place the cylinder into the bottom of the 1-1/2" cut hole, and push the cylinder up through the hole. As you are holding the cylinder, place the coil spring and platform cup into the lower A-arm. Make sure that the cylinder shaft extends into the center of the platform cup.

Cut a hole 1-1/2" towards the middle of right side of the trunk. Push long hydraulic hose into the hole, and run hose up to the engine area.

Tie the long hydraulic hose with heavy duty wire or tie wraps so that the hose will not drag on the ground. Tighten all other hoses together with a Crescent wrench.

Drill 2 holes which are 5" apart on the right side of the trunk and 5" from the quarter panel, so 9/16" bolts will hold the hydraulic pump in place. Tighten bolts. Drill holes for the solenoid and connect all wires.

Run the switch cord from inside the car, through the back seat or under the carpet, to the trunk. Connect the switch cord to the solenoid, dump valve, and battery. Tape all areas of stripped wire with electrical tape.

Connect the batteries in a series circuit. Be sure to have a good ground on the first battery. Now, you are ready to test the system.

After the car has been raised and lowered one time, look at the upper A-arms to make certain they are not hitting the hydraulic cylinders. If the cylinders are hitting the A-arms, cut a small amount of metal off the A-arm until the A-arm clears the cylinder.

DO'S AND DON'TS OF HYDRAULICS

1. Charge the batteries one at a time (depending upon the charger).
2. Always click the switch to go up or down. Never hold your finger on the switch.
3. Use new 30-weight motor oil to fill the pump.
4. Check oil level once a week. If the pump squeals, the pump is low on oil.
5. When the solenoids click, the batteries are dead, or there is a bad cell in one of the batteries. Don't hit the switch up, because the solenoids will stick.
6. Don't run more than 3 batteries for 1 pump, and not more than 4 batteries for 2 pumps.
7. Do not tamper with any cables or wires in the trunk.
8. Check to see that the pumps are tightly grounded.
9. Make sure all cables are tight on battery terminals.
10. Fill the pump when the car is down.
11. Keep the fitting on the return tank loose, so air can come out.

by Ron Ceridono

LATE REARS

FOR FIFTIES FORDS

There's this guy we know, named Rocky Lau, who is the product of a privileged upbringing. No, his dad wasn't rich, it was better than that, his pop ran a wrecking yard!

Access to all those treasures, parts and pieces that the uninformed refer to as junk, has made Rocky an astute assembler of things automotive. He has discovered that installing late parts in early cars is often a low dollar, high ingenuity proposition. As Rocky puts it, "I believe there is nothing wrong with using used parts. It gives you experience when you remove a part, clean it, inspect it and reinstall it." Right-on Rocky.

When it came for a rearend swap to update the mechanicals of his '54 Ford custom, Rocky armed himself with a tape measure and hit the boneyard to see what would fit. After looking at and measuring some likely candidates, a '72 Maverick unit was found that fit the bill in width. The lug bolt pattern is the same as that of the '54, and in addition the later rear is much stronger and has far better brakes than the original. The one problem was the spring pads, the Maverick mounts were 1 inch too wide to bolt to the '54 Ford springs. The normal solution in this situation is to cut the mounts off with a torch, move them the required amount and weld them back on. But Rocky felt that the slight difference in the location was not worth the trouble to cut and reweld the pads. In addition, welding new mounts on an axle housing can cause warpage if you're not careful, resulting in premature bearing failure.

Rocky's method of mounting the rearend consisted of a 1/2-inch steel adapter/spacer with a locating pin and hole to allow the Maverick housing to bolt to the original Ford springs, without moving the mounting pads (see diagram).

The neat part of mounting a rearend this way, in addition to its simplicity, is that the adapter can also be a lowering block. By using thicker material, or perhaps square or rectangular heavy-wall tubing, different width spring/pad combinations can be made to fit.

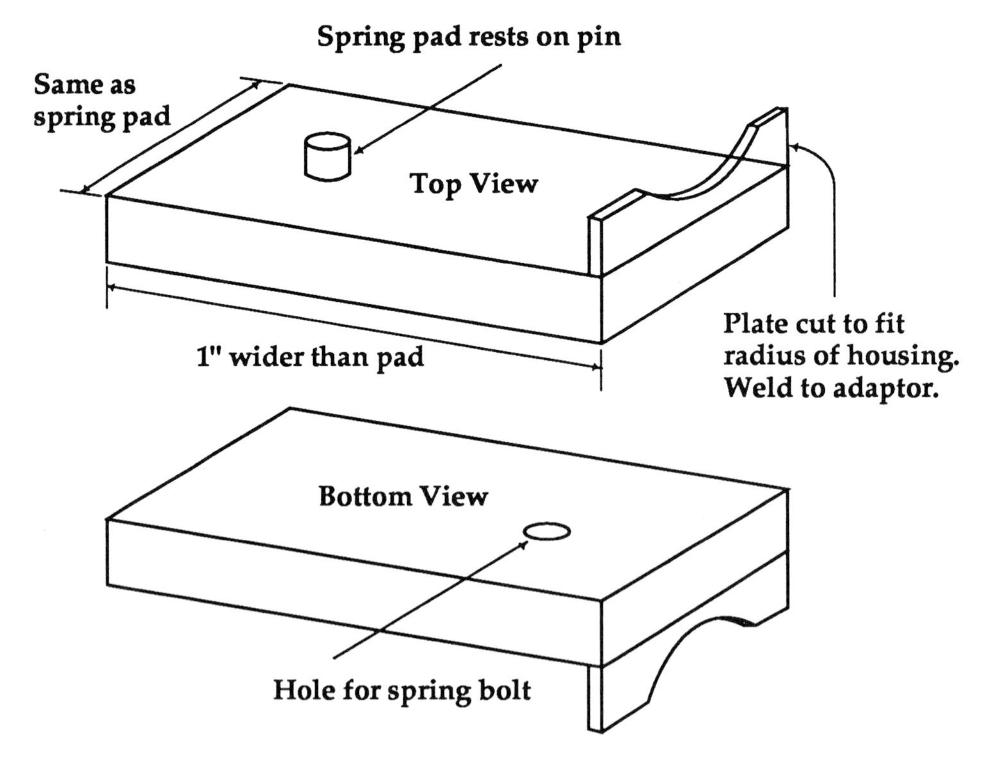

Grab your tape measure and have at it.

"The Ultimate Wall Hanging"

FOR
HOME - WORK - BUSINESS

1948 FORD

Complete details on the 1948 Ford are listed below.

1950 Mercury

Un-Painted Front **$249.95**
a. Bumper Stock .... $ 59.95
b. Bumper Briz Type .... $ 39.95
c. Headlights (per pair) .... $ 29.95
d. Parking Lights w/Grill Assembly .... $ 89.95

Optional Items

Painted Front **$499.95**
a. Bumper Stock or Briz Type, Fiberglass
b. Headlights, Fiberglass
c. Grill and Parking Lights, Fiberglass

1957 T-Bird

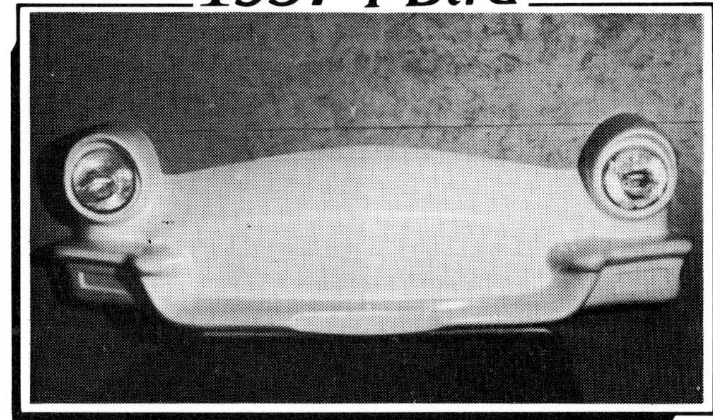

Un-Painted Front **$224.95**

Painted Front **$399.95**
a. Choice of Color, Ready to Hang on the Wall

The aura and magic of the fabulous forties and fifties is continued in this limited edition of automotive artwork.

1948 Ford (Pictured above)

Un-Painted Front **$199.95**
a. Bumper Stock Plain .... $ 39.95
b. Bumper Briz Type .... $ 39.95
c. Headlights .... $ 29.95
d. Parking Lights .... $ 15.95

Optional Items

Painted Front **$399.95**
a. Bumper Stock or Briz Type in Fiberglass
b. Headlights - Fiberglass
c. Parking Lights - Fiberglass

Ball's Hot Rod Parts now offers a fantastic original size fiberglass recreation of some of the most famous Fords and Mercurys ever built by Henry Ford.

Coming Soon!
1957 Chevy
1941 Willy's
1940 Ford - Deluxe

VISA-MASTERCARD Welcomed
Dealer Inquires Welcome

STEERING U-JOINTS

Courtesy of Gerald Zordan and Sharon Marinelli
Borgeson Universal Company

Steering is not normally thought of as a method to increase a car's performance. But if the existing steering system doesn't allow for things like a larger engine or headers, moving the steering can indirectly increase power by making such changes possible.

When building a car, steering can be one of the last items put in. Designing your own steering system allows for the flexibility necessary to get around all the obstacles in the way of a direct line between the steering wheel and the rack or box.

STEERING U-JOINTS

Steering universal joints should have no backlash or radial play. Joints built for industrial applications do not need to meet the close tolerances necessary for steering. Joints built for steering can be loosely classified into two basic categories, those with needle bearings and those without needle bearings. Each has uses to which it is better suited. Nearly all production passenger vehicles use needle bearing joints because they are long-lasting and maintenance free. Where a high strength to weight ratio is important, and longevity is not a major consideration (such as racing applications), non-needle bearing joints should be considered.

Steering joints are made in different sizes. The size (outside diameter) and the material used in making the joint determine the strength of the joint, which in turn determines the acceptable applications in which the joint can be used. A joint made for a lightweight rear-engine vehicle will not perform well under the weight and stress of a V8 over the front wheels. In other words, IT MAY BREAK.

Carbon steel joints, while cheaper to manufacture, have considerably lower strength than heat treated alloy steel joints of the same size. Therefore, a carbon steel joint adequate to do the job of an alloy steel joint has to be larger in diameter to be of comparable strength. Carbon steel joints can usually be identified by the necessity of a hardened steel sleeve being inserted in the yoke around the pins. This sleeved style joint should not be confused with a needle bearing unit, which is closed on the end to prevent grease from escaping and dirt from contaminating the needles in the bearing.

Operating angles for joints should not exceed 30 degrees. Some industrial joints can be operated up to 45 degrees, although the catalogs from these manufacturers almost always note that joints should not be used at an angle greater that 30 degrees, except in a low-stress application. Steering is certainly not low-stress. Consider the following. Driving down the road at 60 miles per hour, taking a wide curve, and striking a pothole. This puts a tremendous load on the steering components, and a joint operating at an excessive angle may not take the strain.

If you have any questions concerning the right size or design of joint to use, consult with the manufacturer of the joint as to recommendations and ultimate torque figures for that product.

STEERING ASSEMBLY DESIGN

Safety should be the first consideration when designing any steering component. The federal government, in one of its better moves, said the steering column should not impale the driver in an accident. Many people are alive today because of that directive.

The way this is usually accomplished is by building some sort of collapsibility into the steering system. Many columns on production passenger cars collapse within themselves. Because this gets complicated and is expensive to do, another method of building in collapsibility is by using two universal joints. Position these joints so a direct hit on the front of the vehicle will not allow the output shaft to transfer the energy of the impact in a straight line back to the driver.

SHAFTS

First some terminology. From the steering wheel to the first joint is the input shaft. The shaft between two joints is an intermediate shaft. And the shaft from the final joint to the steering box or rack is the output shaft.

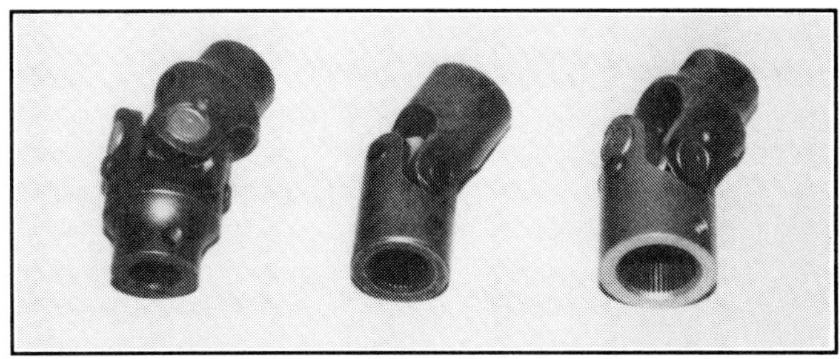

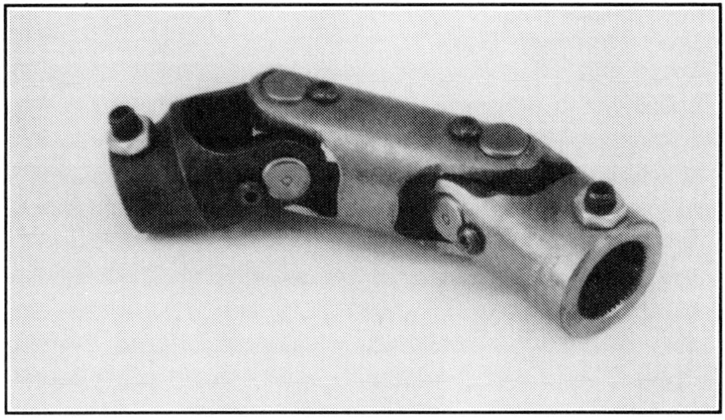

For angles in excess of 30 degrees, double universal joints are available.

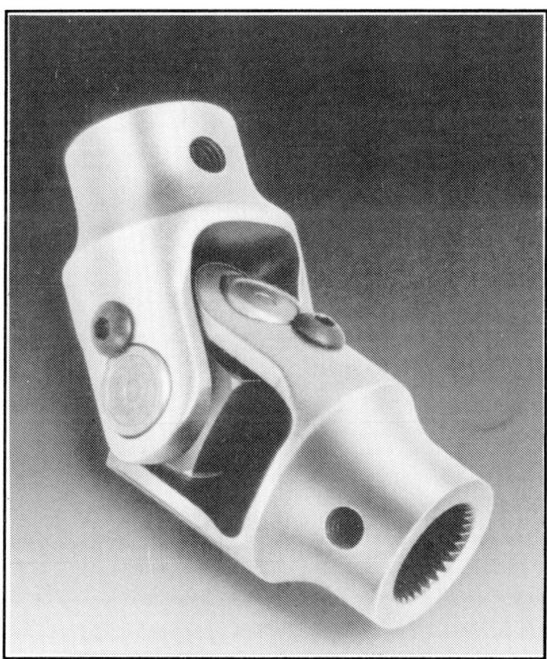

The two best methods of securing the universal joints to the shaft are by splines...

The shafts should be arranged such that no joint has a greater angle of operation than 30 degrees, as stated earlier. If the angle must be greater than 30 degrees, do not alter the joint to accommodate that angle. This weakens the joint, both by reducing the amount of steel available to support the system, and also by dramatically increasing the stress on the joint itself. A joint can transmit 100% of its capacity in a straight line. At 30 degrees, it can only transmit 85% of its capacity. At 45 degrees, it loses an additional 15% of its strength. For angles greater than 30 degrees, use two joints back-to-back, or a double joint (readily available from most manufacturers).

Joints on an intermediate shaft should be aligned so the ears of the yokes on the shaft are in line, not 90 degrees out of phase. In many applications, a third or fourth joint may be used (and occasionally more). The shaft between these joints is also an intermediate shaft, and the joints should be positioned accordingly.

If only one or two joints are being used, a support bearing is not necessary. For three or more joints, support bearings must be used so that there are never more than two joints in a row without a support bearing between them.

A rod end bearing is commonly used in racing as a support for steering shafts, however a pillow block is a stronger support for the shaft. The bearing should be of sufficient size to prevent failure in high-stress situations. It should be mounted in a rigid, solid, strong area, not through a sheetmetal section of the body.

Shaft size is an often overlooked variable in steering. A 3/4-inch diameter low-carbon steel solid shaft will shear at about 339 ft/lbs of torque. This is more than most men are capable of applying with a 14-inch-diameter steering wheel. Using 3/16-inch wall, 3/4-inch-diameter tubing will reduce weight by 25%, while only reducing strength by 6% to 12%. A 5/8-inch-diameter shaft will reduce weight by 30%, however, this will rip apart at only 195 ft/lbs of torque, a strength reduction of 42%. The average man is capable of putting this much pressure on the shaft under extreme situations.

Just to keep things in perspective, remember this

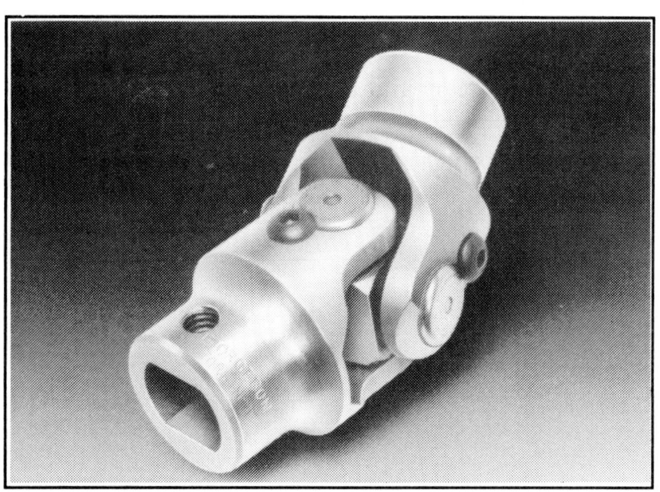

...or as many manufacturers now use, the double-D.

fact: Ford Escorts, generally not considered to be a performance car, use a 3/4-inch shaft. These cars were designed with nowhere near as much weight over the front wheels as a V8-powered rod.

Design your steering as if your life, or your customer's life, depends on it. Remember, steering and brakes are the two most important things on a car.

ATTACHING JOINTS TO THE SHAFT

Four methods of transmitting torque from the universal joints to shafts are commonly used today. They include splines, pinning, welding and keying.

SPLINES: Detroit uses irregularly shaped shafts such as splines or a "double-D" configuration, and inserts them into an irregularly shaped hole with practically no play, and then secures it by staking or clamping. Since steering failures are practically unheard of in modern production cars, one should strongly consider this method as having significant merit.

PINNING: Commonly done by drilling and inserting either two 3/16-inch-diameter roll pins at 90 degrees from each other, or with one 3/16-inch-diameter hardened shear pin. Extreme care must be taken in drilling the holes so they are not oversized and will not allow play to develop or the pins to fall out. A 3/16-inch-diameter drilled hole will weaken a 3/4-inch shaft by almost 30%. It will weaken a 5/8-inch-diameter shaft by 38%. This method may be unacceptable for a 5/8-inch-diameter shaft. Some people use a 1/4-inch-diameter aircraft quality bolt. A hole for a 1/4-inch-diameter bolt will weaken a 3/4-inch-diameter shaft by 42%.

Inserting pins, traditionally done with a hammer, can damage the steering joints if the assembly is not supported at the back side of the point of contact by the inserting tool. Allowing the impact of insertion to go through the working parts of the joint can destroy the integrity of the joint, and cause premature failure.

WELDING: A method commonly used in racing, welding may be illegal for highway use in some states, and is a method the authors do not recommend. This method has several drawbacks. Needle bearing joints can have the grease melted out of them at relatively low temperatures (as low as 140 degrees F). This will cause premature failure of the joint. If you can't hold your hand on the working parts (center) of the joint immediately after welding, the bearings may have been overheated. Non-needle bearing joints, while they can tolerate higher welding temperatures, can also suffer. Higher temperatures can distort the steel and cause the joint to bind, and ultimately lead to failure of the system.

Improper grounding during welding can also cause joint failure. Attaching the ground of a welder so the current passes through the pins and yokes of a joint can cause the pins to be welded to the yokes. If this happens, the joint will fail. If you choose to weld the joint to the shaft, be sure all the pins move freely as they were originally intended to do.

Another problem with welding is the overheating of a joint made of heat-treated alloy steel. The heating and cooling process can cause some of the joint's strength to be lost, thus shortening its life. Rapid cooling of a weld can cause cracks in both the joint and the weld, leading to failure of the system. Welding makes a permanent connection which does not allow for disassembly of the system should repairs need to be made, either to the system itself or to any other component of the car which may be blocked by the steering shaft. With the possibility of all of these problems, welding should only be done by a qualified welder in whose care you can trust someone's life.

KEYING: Using a key, as is done in many industrial applications, can transmit power effectively from the shaft to the joint. A key, however, is not suitable to take sudden shock. Sudden impact (such as from a pothole or accident) can distort the key or the shaft keyway. This can cause play to develop in the system.

SETSCREWS: These should never be used to prevent the joint from turning on a shaft. They should only be used as a method to prevent the shaft from disengaging from the joint.

MAINTENANCE

Today's cars from Detroit and Japan have relatively maintenance-free steering systems. These companies have large engineering staffs to make sure that nothing can go wrong, yet the word "recall" has a well understood meaning when used by the major car manufacturers.

When making something without the help from thousands of engineers and millions of cars of experience, the likelihood of a malfunction is much greater.

Take the time to check the steering often. Be sure play is not developing. Turn the steering wheel to see how far it moves before the wheels react. Replace worn parts as necessary. Lubricate non-needle bearing joints. Boots can hold grease in and keep dirt out and prolong the life of this type of joint.

Inspect the connections between the joints and shafts. Be sure any clamp screws are not loose, and check for pins that can fall out. Welds are difficult to inspect, but look for cracks.

The benefits of moving the steering, or building your own, are well worth the few minutes needed for inspection to keep your steering working as it should. Besides, regular inspection may save a life.

ALIGNMENT

Once you have the suspension system installed, all would seem well and good. It is ... almost! Now, you must make the wheels roll true, and do what they should do in a turn.

Leave the final wheel alignment to the professional shop. However, there is a lot of alignment that you can do at home to get things at least in the ballpark. Since most projects are a long time from first movement to final drive-away, this initial alignment will help things considerably.

A note here: If you have an independent rear suspension, it is vital that this unit be aligned by the professional. Such systems have a lot to do with how the vehicle will handle, and they are not simply set with the wheels parallel to the chassis and vertical to the ground.

Rearend alignment is supposed to be right on the money if you have measured carefully when attaching all the mounting brackets. Quite often it is. But to find out, measure diagonally from the leading edge of the rearend housing at the outer brake backing plate flange, to some known point on the opposite frame rail, well forward. Do this on each side to the opposite rail. If the frame has measured square, this will give exact true to the rearend. Measure across the chassis from the rearend housing flange to make absolutely sure the rearend is centered under the frame. If the rearend must be moved, now is the time to do it.

Front end alignment is similar. The axle at the spindle should measure identical on a diagonal to a frame point. Adjustment at the radius rod mount(s) or the A-arm mounts will bring this into "square." Now, lay the kingpin inclination backward until there is about 5 degrees of caster in the spindle. The wishbone or 4-bar set-up can be adjusted, and shims are available to be placed between A-arm mounts and the A-arm itself. Since there are a number of different adjustments in the A-arm system, ask the front end professional for adjustment points with your particular system.

Wheel camber, or the amount the wheel leans in at the bottom versus the top, is not important at this time, although a sighting down the wheel line from in front should show the bottom tilted in slightly from the top.

If you have decent caster, and the camber is usable, then the only other factor is toe-in/toe-out. Measure across the wheels from one side to the other, using a tire sidewall or tread mid-point as reference. Generally speaking, at this stage, something like 3/8" toe-in will work. That is, the measurement across the front of the tires will be closer together by 3/8" than measuring the same place at the rear of the tires. If you do not have a toe-in or toe-out factor in the front end, you'll feel a lot of shimmy in the steering.

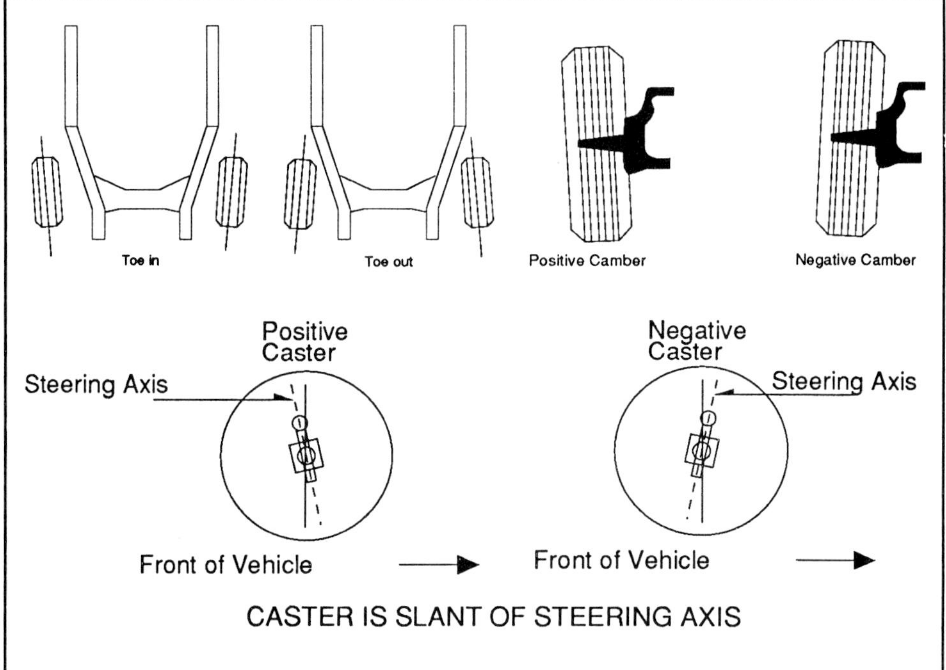

CASTER IS SLANT OF STEERING AXIS

Trying to measure the rear of the tires can be a problem (as in the front if there is sheetmetal in the way). A quick solution is to tape plumb bob weights to the center of the tire, front and rear, then measure from these (near the floor).

Sometimes, no amount of toe-in work seems to remove shimmy from a front end. Try as much as 1/2" toe-out. This often cures the problem. Of course, shimmy can also be caused by excessive play in the kingpin bushings, or excess play in the tie rod ends. Also check the steering gearbox for wear.

THE

ACKERMAN

PRINCIPLE

Consider the car as it turns in a constant circle. The inside front wheel is turning a smaller-radius circle than the outside front wheel. Something must be done in the front steering mechanism to allow this to happen. This something is called the Ackerman principle, and if you look at a set of spindles you see that the steering arms on the spindles have the tie rod end holes closer to the center of the car than the kingpin holes.

If you draw a line from the center of the kingpin to the center of the rear axle, the line should pass directly through the steering arm rod end holes. If not, the car does not have perfect Ackerman effect. Current mass production technology is beginning to ignore this principle somewhat. Some new cars with rack and pinion steering ahead of the front axle centerline actually have the outer tire turning tighter than the inner wheel. Tire technology is offsetting some of this problem and so is wheel offset, but for our purposes, stick with making a pure Ackerman effect on your rod.

If it is necessary to bend a spindle steering arm for any reason, be sure and set it so that the Ackerman check line described is obtained. Sometimes, when a crossleaf spring is used on a frame with a suicide spring perch, the tie rod runs into frame interference. Rodders have cured the problem by reversing the spindles side-for-side. This puts the tie rod in front of the axle. And the Ackerman goes out the window, resulting in very poor turning control at higher speeds. If there were enough room, the spindles could be heated and bent outward so that the tie rod holes would again line up for the Ackerman check. Not really conceivable, so better to find another way and keep the tie rod behind the axle.

One method is to mount a rack and pinion steering gear directly to the solid axle. This is usually a simple matter of two sturdy brackets between the rack and pinion unit and the axle. This creates a problem with the steering shaft, however. As the axle travels up and down, the effective length of the steering shaft changes. Some new cars use spline sections in the steering shaft, and rodders cure the variable-length problem this way.

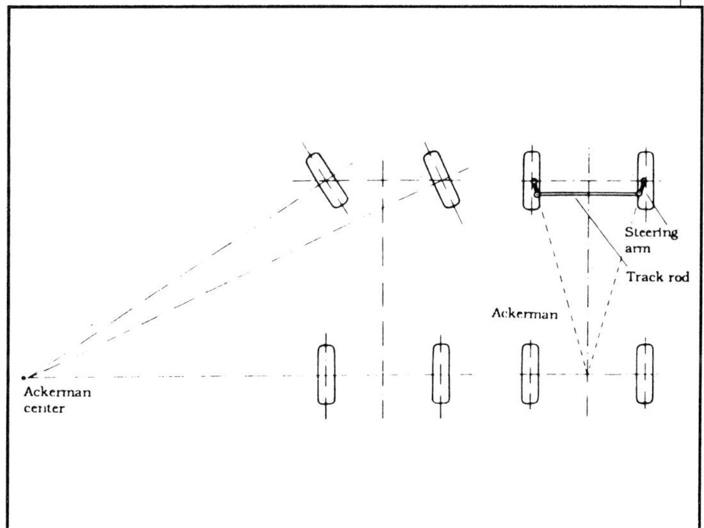

SCRUB LINE

One of the more important checks in any vehicle safety test would be that of the scrub line. That is, any part of the chassis/body that hangs below the wheel's diameter. Unfortunately, a large percentage of hot rod builders violate this basic safety tenet, with both early and late model vehicles.

The reason for not wanting to have anything hanging lower than the bottom wheel lip is obvious. Given a flat tire, the offending part(s) can cause serious problems. One example would be a steering pitman arm that is too low. It digs into the pavement and the car can go out of control instantly. A bracket or such grinding against the pavement sends up sparks and a fire can result.

While it is true that a flat tire seldom lets the wheel rest on the road surface, good building sense says never take chances. To check for scrub line violations, have a buddy hold the end of a long string while you hold the other end. Start by running the string between the bottom edges of the two front wheels. Then move the string diagonally from one front wheel to the opposite rear wheel. If there seems to be something hanging too low, it is worth a further check, and repair if necessary.

by Warren Gilliland

BRAKES

During the 1930s and '40s, Ford automobiles were undergoing tremendous advances in styling, performance and comfort. In a matter of 10 years, (1940-1949), the automobiles coming down the assembly line at Ford looked totally transformed. The average person with little knowledge of automobiles would probably be unable to identify the '40 Ford and the '49 Ford as being made by the same company. Advertising was stressing the three areas mentioned as the reasons to purchase a new car. Handsome men and beautiful women driving convertibles painted a lovely picture of taking a drive just for the pure pleasure of it.

Unfortunately, despite the fact that the car had substantially improved from a comfort and design standpoint, the brake system was virtually unchanged. The '49 Ford and, in fact, the '50 and '51 Ford as well, still had most of the elements of the early Ford hydraulic brake system. Finally, in 1952, a complete redesign of the brake system took place, and the components found in a '52 were the same until they were again changed for the '57 models.

The key difference between the early and later shoebox Fords is that the early Fords still have the master cylinder located under the floorboard. It wasn't until the '52 Ford came out that the master cylinder finally was moved up to the firewall. As insignificant as this may seem, this constituted a major improvement in brake safety for a couple of very important reasons. First, under the floorboard, bolted to the frame, the master cylinder was exposed to all of the elements of the weather. Since the master cylinder is made from cast iron, it is very vulnerable to moisture, which in turn causes rust. Once the bore of a master cylinder begins to rust, it also begins to leak. Loss of brake fluid will result in total loss of brakes. Since the master cylinder was also out of sight under the floorboards, it was frequently overlooked during maintenance checks, further increasing the likelihood of eventual brake loss.

Another serious problem with having the master cylinder down low is that it is much more likely to be damaged if the car runs over anything in the roadway high enough to scrape the bottom of the car. Forty years ago, the condition of the roads in the United States was considerably rougher and harder to navigate than today. In virtually every part of the country, much of the nation's roadways were still dirt

Pedal Ratio Diagram

and crushed rock. Potholes and uneven surfaces were encountered on a regular basis. Components on the bottom side of an automobile were replaced with much greater frequency than today. If you were purchasing a used car 40 years ago, you never did so without checking the undercarriage to see how the car had been used. If you failed to do so, the result could be a major out of pocket expense for repairs. Brake failure was so frequent that the device we have on our automobile today, known as a "parking brake" was much better known back then as an "emergency brake."

The 1949 Ford did have one key improvement over earlier brake systems. The earlier cars had a fixed-pin brake system, which simply means that the brake shoes were actuated by the wheel cylinder pushing out on the top of the shoes, the front shoe moving forward, and the rear shoe moving backward independently of each other. There was a fixed-pin at the bottom of the brake system, and the shoes expand out against the drum to create the friction necessary to slow the car. In order to achieve the best brake in this type of system, it is extremely important to keep the brakes in proper adjustment. Since self-adjusters were not available in those years, this meant having your brakes adjusted every few thousand miles, usually with every oil change. In 1949, this was changed to the "floating" shoe style of brake. In this type of system, the wheel cylinder still pushes out on both shoes, but the bottom of the shoes are allowed to float freely with each other. As the front shoe moves out to contact the drum, the bottom of the front shoe moves to the rear. Since the front and rear shoes are now moving together, this forces the rear shoe into the drum from both the bottom and the top (don't forget, the wheel cylinder is still pushing on the top of the shoe). The rotation of the brake drum causes the shoe to be pulled into the drum and, in essence, creates a more rapid increase in brake torque, making the brake feel "self energized."

This basically means that you get more brake than you would have gotten from the same pedal force with the old system. The effort required to stop the car is reduced, and the feeling is more like a power brake. This feature of self energized brakes is why the brake system on an older car is so unreliable in wet weather. When rain enters the brake drum, it reduces the friction between the shoe and drum, requiring more pedal effort to get the same stopping ability as you would have had in dry weather. You may have noticed that late model cars equipped with disc brakes don't loose any significant amount of stopping ability in wet weather. This is because the system doesn't rely on being self-energized to create the friction necessary to stop the car.

In 1951, Ford increased the width of the front shoes, and that added to the stopping ability of the car. The wider shoes not only helped the car stop better in repetitive stops, but increased the life of the lining as well. As driving speeds increase, more frontal weight transfer takes place during stops, which increases the loads that the front shoes see. The increased surface area of the wider shoes quickly took care of the problem that the extra heat being generated caused the system.

When the master cylinder was moved to the firewall in 1952, the entire pedal assembly was inside the passenger compartment, and the master cylinder was up underneath the hood in a much safer location. Since the master cylinder was now much easier to get to, and could be seen during normal inspection of the oil and water level, it was more common to check the brake fluid level as well. If the master cylinder did begin to leak, it would now do so down the firewall, making it easier for an alert mechanic to catch before total brake failure resulted. This was a major help in

Typical Tandem Master Cylinder

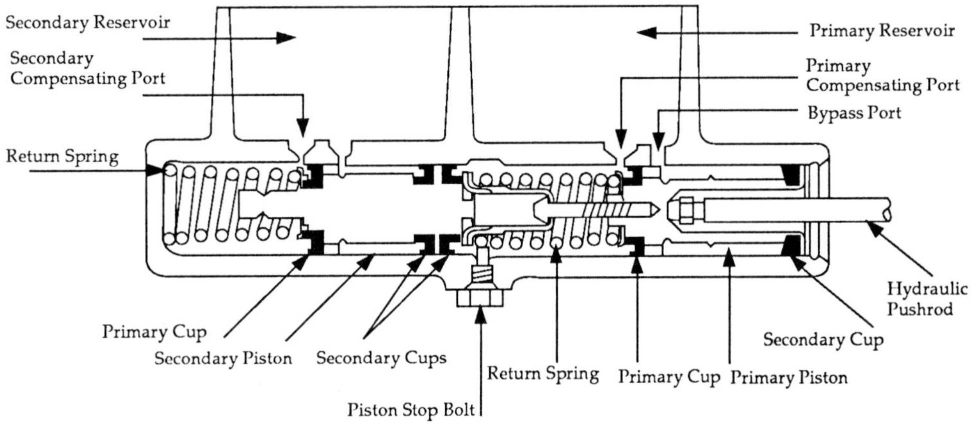

Secondary Reservoir
Secondary Compensating Port
Return Spring
Primary Reservoir
Primary Compensating Port
Bypass Port
Hydraulic Pushrod
Primary Cup
Secondary Piston
Secondary Cups
Piston Stop Bolt
Return Spring
Primary Cup
Primary Piston
Secondary Cup

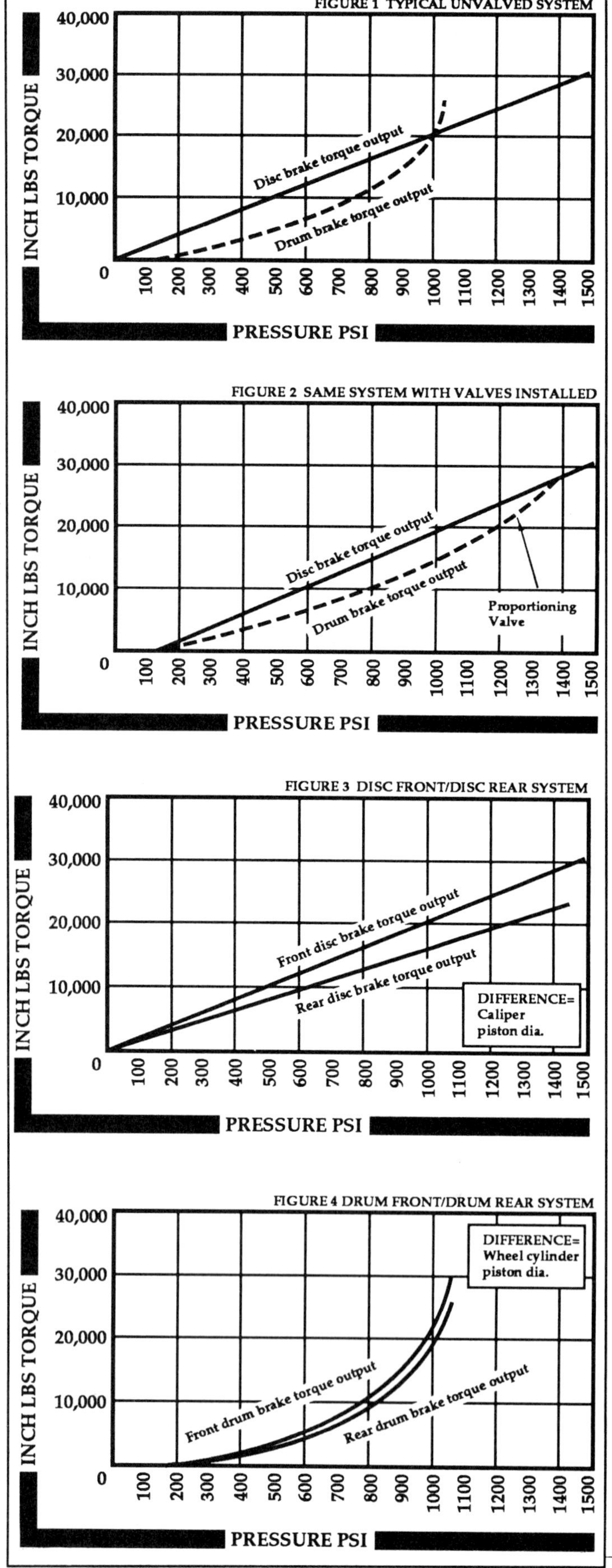

reducing brake failures during this era.

In 1954, Ford made the last significant brake improvement until the advent of the disc brake. This was the vacuum power brake, which was mounted on the left fenderwell under the hood. Basically, engine vacuum is supplied to a large diaphragm. When the brake is depressed, the line pressure is magnified by the power booster, resulting in the same stopping ability with less effort. These improvements made it easier for people who lacked physical strength to drive an automobile comfortably. It also increased the stopping ability of the car in emergency situations. This system was so effective that it was used for many years, virtually unchanged, with the exception that the power unit was moved from the left fenderwell to behind the master cylinder a few years later.

Now that you know the history of the changes in the shoebox Ford brake system, let's explore how the brake system works, and what each component in the brake system is supposed to do. You will find this extremely helpful when we start looking at the easiest ways to improve the stopping ability of your car.

BRAKE PEDAL AND MASTER CYLINDER

The brake pedal and master cylinder work together to create the system operating pressure, so we will discuss them as a unit. The master cylinder must be large enough to supply a sufficient amount of fluid to actuate all of the calipers and wheel cylinders that will be feeding from it. This must happen in approximately half of its available stroke, but not any more than 2/3 of a stroke. The second, and most often overlooked, requirement of the master cylinder is that it must have a reservoir large enough to accommodate sufficient reserve fluid to allow for the displacement required for total wear of the lining, and subsequent piston movement. When the linings and/or pads are in a totally worn condition, the reservoir should still have approximately 25% of its total reserve in the tank for a safety factor. Failure to make sure that your system meets this requirement could result in a partial or total system failure.

The brake pedal is designed to be in a comfortable position for use, but more important, since it is a lever, it must be of a proper ratio to give the best results. The pedal ratio refers to the measurement from the pivot point to the master cylinder pushrod point, when compared to the measurement from the pivot point to the center of the pedal. If the first measurement "A" (3 inches) is divided into the second measurement "B" (12 inches), the ratio is 12 ÷ 3 = 4, or 4:1. If you wish to increase leverage, it can be done by either lengthening the "B" dimension or decreasing the "A" dimension. If you decrease the "A" dimension to 2 inches by drilling a new hole for the pushrod, the formula would now be 12 ÷ 2 = 6, or 6:1. If you used the same amount of pushing effort from your foot, you will now have increased the brake pressure by 50% in the second equation. The problem is that your pushrod will be pushing into the bore on an angle. It will be necessary to relocate the master cylinder closer to the pivot point or premature wear to the master cylinder bore will result. Additionally, this will increase the pedal travel by 50%, creating an uncomfortable and slow reacting pedal. Small changes make a large difference. Be careful to consider all factors before making any modifications in this area.

There is a formula that clearly shows the relationship of the master cylinder bore size and the pedal ratio. If you use it and plug in different values, you will be able to determine what your system is doing and whether it is in proper proportion. This will also allow you to explore the best way of making system modifications that achieve all your goals. Here it is: Input pressure X pedal ratio ÷ area of the master cylinder = line pressure in PSI (pounds per square inch).

Let's try the formula. Input pressure refers to the force from your foot on the pedal. Most people can push 150 pounds with effort, under a maximum braking situation, and it is a good number to use here. If you want to test, use a bathroom scale in your car on top of the brake pedal, and see what you can do. We don't want the brakes too touchy, or they will lock easily, which will hurt stopping distance. The ideal pedal ratio of a manual brake (no power booster) is about 6:1. The surface area of the master cylinder refers to the surface area of the bore size. Example: 7/8" diameter = .60 square inches; 1" diameter = .79 square inches; 1-1/8" diameter = 1.01 square inches. Okay, now let's plug in some numbers.

150 (input pressure) X 6 (pedal ratio) ÷ 1.01 (surface area of a 1-1/8" diameter bore master cylinder) = 891 (PSI line pressure). A number of around 900 PSI line pressure, when used to supply adequate components (calipers or drum brakes) will result in a proper operating effort for most cars. On heavier cars, you may want the system pressure a little higher, and on lighter cars you may want it a little lower. But the example supplied here helps you understand the relationship of the numbers. Never exceed the maximum suggested operating pressure of the system. This is usually 1200 PSI, but always check with the manufacturer of the components first.

If you increase the input pressure or pedal ratio, the PSI increases. If the master cylinder bore size increased, the PSI drops. The system pressure can be increased by decreasing the bore size of the master, but don't forget about its primary duties mentioned earlier. This is only an option if the other criteria can be met.

If your system has a problem of simply too high an effort to make the brakes work, your entire problem may lie with what you have just learned about developing line pressure. It doesn't make any difference how big the calipers or drum brakes are, if they don't receive sufficient pressure they will not do what they are capable of doing.

It is not only important that the master cylinder create pressure, it is also important that it relieve the pressure when you take your foot off the brake pedal. If the master cylinder has an internal residual valve (made to operate with drum brakes by maintaining about 10 pounds of line pressure, not enough to overcome the drum return springs, but helps to substantially reduce pedal travel), it will not release the pressure sufficiently for disc brake calipers, and a serious drag will result, causing extreme wear. This, in turn, causes overheating of the brake system and unnecessarily high load on the drivetrain. If you do not allow the brake pedal to return all the way, it is possible that the piston inside the master will not return far enough to open the pressure compensating port. This would also result in serious brake drag.

Because the 1949-'51 Ford cars have the master cylinder located under the floorboards, this places the master cylinder below the height of the calipers and drum brake pistons. This causes the fluid to attempt to drain back to the master cylinder. Since many modern master cylinders do not have an internal residual valve, this could result in a condition where the next stroke of the brake pedal will not be able to supply enough fluid to actuate the pistons, causing a temporary brake failure. This condition would be a potential problem if you are going to adapt disc brakes somewhere in the system, or if you intended to replace your stock master cylinder with one from a newer vehicle. If your car has drum brakes all around and a drum brake master with an internal residual valve, you will not encounter a problem. If you do not have an internal residual valve, follow this simple rule: Install a 2-pound external inline residual valve for disc brake calipers, and a 10-pound inline residual valve for drum brakes. Since the drum brakes have large return springs pulling the shoes back, they are more than capable of pulling the shoes back in spite of the valve. The disc brake caliper, however, has nothing pulling its piston back after application, and residual pressure will cause it to remain applied. The 2-pound valve creates so little force, no drag problem will be created.

One final note of caution: If your master is mounted under the floor, make sure to keep the exhaust system at least a foot away from the master cylinder. Failure to do so will transfer heat to the fluid, causing the fluid to boil, potentially resulting in brake failure. If you must run the exhaust near the master, use a deflection shield or insulation material to prevent the heat transfer.

At the present time, a wide variety of speed shops are recommending the old 1-1/8" bore fruit jar master cylinder that was commonly found on Fords of the early '60s, including the Mustang. This master cylinder, although it has sufficient volume, is not a good choice as it is only a single-bore master. In the late '60s, the tandem master was developed, which is simply two masters in one. With the old fruit jar master, if there was an internal failure, the brake pedal went to the floor and you had no brakes. On the later tandem cylinders, if an internal seal goes bad, it will probably only result in a loss of half of the system, allowing the car to be stopped with the other two wheels. Since the cylinders found on the later Mustang II are almost as compact, and are tandem as well, they offer a far better choice for a safe system.

CALIPERS AND WHEEL CYLINDERS

The wheel cylinder or caliper (both perform the same function) take the pressure present in the system and turn it into force. The force is the push on the shoes or linings, as the case may be, and is determined by the size of the piston. The larger the piston, the greater the force. On shoebox Fords, the stock drum brakes work reasonably well unless you are considering increasing the weight or installing a modern motor under the hood. The best choice for reasonably and easily improving the brake system in this area of the 1949 and '50 Fords would be to adapt the wider front shoes from the '51 model. The change is simple. Hang the wider shoes on the existing backing plates and use the wider brake drum from the '51 vehicle. No other modifications are necessary.

Another simple improvement is to adapt a larger wheel cylinder. The stock cylinder was 1-1/8" diameter, and by going to a 1-3/16" diameter unit, the larger bore size will result in more force. If you want to modernize the system dramatically, and will be using front disc brakes, the best choice for upgrading the rear will be provided by adapting brakes from the newer Ford 9" rearend. The modifications are minor and the results rewarding.

If you have decided to upgrade to front disc brakes, you have several things to remember. Obviously, you must prepare the master cylinder as mentioned in that section. Don't even begin to worry about the disc brake adaptation until that portion of the job is handled first. Next, you must remember that disc brakes require considerably more wheel clearance, which will require you to make a new wheel selection. If you intend to run stock wheels, be prepared to make major modifications to the new brake components or their mounting location in order to make them fit. If you are intending to use the brake system hard, keep in mind that the later disc brake cars use larger spindles with bigger bearings to deal with the increased loads. Remember that the system is only as strong as the weakest component, so be sure to compare the bearing diameters on your spindles with the new components to make sure you are not creating a weak link in the chain.

BRAKE SHOES AND LININGS

Many people have the mistaken impression that the size of the pad or shoe also determines the amount of brake you get. In the case of a drum brake shoe, the surface area of the shoe is not the factor, but rather the size of the drum that is being used. The surface area of both the pad and shoe determines two things. First, it determines the life of the lining. The larger the surface area, the greater the life. Second, the larger the surface area, the lower the resultant temperature during the same amount of braking. In other words, if you install larger pads, the heat seen by the pad is absorbed by a larger surface and hence the temperature stays lower. In the case of brake lining materials, they have a temperature point at which they lose their coefficient of friction, or ability to grab another material. You also experience this condition when your brakes get wet. You can feel a noticeable change in the braking effectiveness when the coefficient of friction goes down. Once the friction material cools, it will regain its ability to grab effectively again. This is known as fade recovery. In rare cases of overheating, permanent damage to the material may be done. Because using brakes creates heat, you are constantly heat treating the brake material. Overheating it severely could cause permanent damage and require replacement, even though the material is not worn thin. Not only is this heat seen by the brake material, it is also seen by the brake drum or rotor. These components can develop hard spots that will not allow them to be turned flat again, and can cause excessive noise like a grinding sound, even though no metal-to-metal contact has taken place. When resurfacing drums or rotors, hard spots will be apparent as discoloration in the turned surfaces. There really is no serious safety problem. The noise, however, can be terribly annoying.

DRUMS AND ROTORS

The purpose of both the drum and rotor is to remove the heat that is developed during braking. This is done by supplying a suitable friction contact surface and a heat sink to remove it from the friction material. These components function as the drain of a sink, in that they get rid of the heat. Unfortunately, when customizing a brake system, this component is usually too small for the job. One of the major component failures on cars imported from the Orient is that they do not comprehend the harder use put to the brake system by American drivers, and consequently every major Oriental import has had brake problems that are most often caused by totally inadequate rotor size and design. The problem is comparable to running your kitchen faucets full blast and having a drain that is inadequate to get rid of the water filling the sink. Sooner or later, the sink begins to overflow. In the case of a brake system, the result is severe warpage to the components, which causes pedal pulsations or extremely fast friction material wear. If you have ever driven a car that, when decelerated from a high rate of speed, begins the stopping sequence with good stopping power but then seems to fade, you are experiencing lining that is going beyond its thermal ability, probably because the drum or rotor is not removing the heat fast enough.

POWER BOOSTERS

One of the most misunderstood components on the entire car is the power booster. First, since the pre-'52 Fords mount the master under the floor, it is extremely difficult to fit a stock power unit in that location. To get around this problem, the novice and sometimes professional builder as well, opts for a small aftermarket booster. This smaller booster offers a great deal less boost than a stock unit, which defeats the whole purpose of why you are installing a booster in the first place. Actually, you have missed the real question, and that is, "Do I need a power booster for what I'm trying to accomplish?" The answer, in most cases, is no. Most power boosters would not be necessary if more consideration was given to choosing the proper pedal ratio and master cylinder bore size. Since power boosters typically run off of vacuum, they rob engine performance and are subject to not working if the engine dies. If components are chosen wisely, this is one of the first items that can be left out. Of course, if you are dropping a fire-breathing dragon 428SCJ under the hood of a '54, you will probably want to include a booster in the project, only because of the potential speeds you may be preparing to see. If you decide to install a power booster, choose a factory unit that will do some good. Tank size determines the amount of boost, and small tanks often provide very little assist.

PLUMBING

To successfully carry fluid from the master cylinder to each of the wheels is the job of brake lines. Most cars are plumbed with a combination of solid line and flex line. Never use more flex line than is absolutely necessary. Flex line "grows" under pressure, requiring more fluid movement, which makes the pedal feel mushy. Never use large loops, like up over the rear axle. In order for the bleeding operation to be as easy as possible, the lines must be routed so that there are no potential places for the air to be captured. It is a good rule of thumb that once the fluid leaves the master and moves downhill to a point below the wheel cylinder or caliper, the line should not be higher than the outlet port at the cylinder, and certainly not in a sharp arc. If it is, the air will not move easily to the bleed screw and will require the use of a power bleeder to clear the system.

Make sure to use quality steel line, recommended for use as a brake line, and use a double flare to ensure a quality connection. Never use nylon lines, as is seen on go-karts, because this type of line will fatigue over time and is much more prone to being damaged. When you route the lines, stay away from heat sources or other areas that may be subject to damage, and always run the lines where they are least prone to being hit by rocks or other debris. Make sure the lines are tied firmly in place at least every few feet so they will not move or vibrate, which can cause wear and fatigue failure. If you are unsure how to route the lines, look at other cars or check with professionals to ensure all safety conditions are met.

BRAKE SYSTEM VALVING

Even though you may make all of the perfect choices for components to achieve a well balanced system, it will still be necessary to do some fine tuning to really make the system effective under all stopping conditions. The three main valves used in the brake system balancing act are: 1. Residual valve, 2. Metering valve, and 3. Proportioning valve.

As discussed earlier, the residual valve maintains pressure in the brake line even when the brakes are not being used. This valve should always be used with drum brakes. The system requires less pedal travel when residual pressure is present, making the entire system feel firmer and react quicker to the driver. Never use a residual valve with disc brakes, except when the master is lower than the caliper, and even then only use a 2-pound valve.

The metering valve is the most overlooked valve on cars that have been upgraded to disc brakes on the front, as is often done on shoebox Fords. Its purpose is to delay the initial pressure from reaching the front calipers until the pressure moves high enough to overcome the return springs on rear drum brakes (usually around 50 to 100 PSI). When this valve is absent from a disc/drum combination system, the disc brakes do all the work in stopping the car at low speeds. This results in extremely premature wear to the linings. The accompanying graphs indicate that brake pressure from a drum brake does not begin until the system pressure has risen substantially. As soon as the brake pedal is touched, however, the disc brakes begin to apply.

If you have converted the front brakes to disc on any of the years of cars covered by this book, then another valve you will need is the proportioning valve. Its function is to restrict high brake pressures from reaching the brakes it protects. The main purpose for installing this valve is to stop high speed rear brake lock-up. (See Testing The System). Once again refer to the graphs, and notice that the rear drum brake puts out torque on a curve. Even though it takes significant pressure to get a drum brake started, once it starts climbing it climbs rapidly, eventually crossing the disc brake torque output, if left unrestricted. By installing the valve, the rate of increase will not be as rapid and will help stop excessive rear brake torque. If a valve is needed for this purpose on a converted car, it is best to choose an adjustable valve. Kelsey Hayes manufactures an adjustable valve that is perfect for this application.

BRAKE FLUID

A hydraulic brake system will not operate without brake fluid. To a large degree, the level at which it does operate is dependent on the choice of fluid. Among other things, the more important characteristics of brake fluid are high boiling point, consistent viscosity, and good lubricating ability. All brake fluids commonly used in automobiles in the United States are regulated by the Department of Transportation (DOT). The fluid container will have a number such as DOT 3, which refers to the test designation that the fluid meets. DOT 3 and DOT 4 fluids are polyglycol base products and are hygroscopic, which means that they absorb moisture. As the amount of moisture absorbed increases, the boiling point decreases. In a well sealed brake system, these fluids require changing approximately every 1 or 2 years, depending upon the severity of use. Early master cylinders, such as those found on the '49-'54 Fords, will not have a suitable diaphragm capable of expanding to fill the void created by the brake fluid being used in the system to compensate for lining wear, and as a result air becomes present in the reservoir. The air contains moisture and contaminates the fluid. Use of this type of cylinder automatically means that you should change your fluid much more frequently.

This is also why you should never buy brake fluid by the gallon. A half used gallon of brake fluid will result in the other half of the gallon container being occupied by air. This amount of air is sufficient to cause the remaining fluid to become contaminated with an excessive amount of moisture before it is even put in the vehicle. When bleeding the brake system, never reuse any fluid recovered from the system.

Be very careful never to install anything but brake fluid in the system. The seals are made from ethylene propylene rubber, which is not compatible with transmission fluid or motor oil. If either of these fluids is inadvertently placed in the system, it will be necessary to replace all of the seals in the entire system.

There is one area of caution that needs to be noted concerning DOT 3 and DOT 4 fluids. They attack paint, especially if allowed to be in contact for some time. If fluid is spilled on a painted surface, flush with water and wipe dry immediately.

DOT Minimum Boiling Points

	Dry Boiling Point	Wet Boiling Point
DOT 3	401 F	284 F
DOT 4	446 F	311 F
DOT 5	500 F	356 F

DOT 5 brake fluid is more commonly known as silicone fluid. This is because it is primarily a silicone based product. It does not attack paint, and it does not absorb moisture. However, it has some characteristics that, in my opinion, make it unacceptable for use in street driven automobiles. First, the compressibility of silicone fluid is very unstable and changes with temperature. Since the temperature of the fluid in a brake system changes under normal driving conditions, the pedal feel is affected, sometimes radically so. It is also affected by changes in atmospheric pressure (altitude). If you live in high country, don't even consider silicone brake fluid. In the racing industry, where brake systems are subjected to extreme use and temperature, it was discovered long ago to be an unacceptable fluid. The expansion characteristics of the fluid under these conditions would be so severe as to lock up the brakes and not allow movement until sufficient time passed for the fluid temperature to come down. If you experience extreme changes in pedal travel or feel, it is possible that you may have silicone fluid in your system.

TESTING THE SYSTEM

The job of building a safe brake system cannot be considered complete until it has been tested and proven to perform under conditions similar to those it will see when actually driven. Go to a safe area to perform some deceleration testing, subjecting the car to varying types of stops to check for feel, safety, and balanced braking under a variety of conditions, including both easy and hard deceleration. Begin with easy stops from low speeds, to check pedal height and to be sure no side-to-side pulling is present. Once satisfied that there are no major errors, make several stops from about 30 mph in rapid succession, to check for brake fade. You should be able to make at least 3 stops from 30 mph with no noticeable change in braking characteristics. Now from the same speed, attempt to make a hard deceleration and see if there is either front or rear wheel lockup. If so, the system is not yet properly balanced and needs more fine tuning. This should be checked up to full highway speed, prior to actually driving on the public streets. Never drive any highly modified car until it has been thoroughly tested for handling characteristics.

For the cars we are covering in this book, the most likely source of out of balance will probably be rear wheel lock-up, since as the driving speed increases and the demand for deceleration increases, weight will shift to the front wheels, creating less weight on the rear wheels. As the weight on the wheels decreases, the ability of the wheel to lock up gets easier. Watch carefully for this condition. The reverse could end up being true if you modify your car by installing oversize rear tires (referring to height, not width). The taller the tires, the longer the lever arm, and the more torque required to lock the wheel. You may use a proportioning valve to correct an out of balance condition, but it is wise to do as much as possible with the basic choice of modified components first.

Don't forget to check the vehicle for hard deceleration from higher speeds. As you can see in the graphs, the rear drum brake torque output increases rapidly. If the rear brakes lock up before the fronts, the rear of your car will be attempting to travel faster than the front, and the car will spin out. Do not drive a car that has this condition, because it can be extremely dangerous.

NOTES FOR THE RESTORER

If you are only interested in restoring your vintage Ford, here are a few special items to watch out for.

1. Watch for signs of leakage from both the master cylinder and wheel cylinders. Because these components are frequently overlooked during maintenance, the possibility is high that the brake fluid is old, contains a great deal of moisture, and has

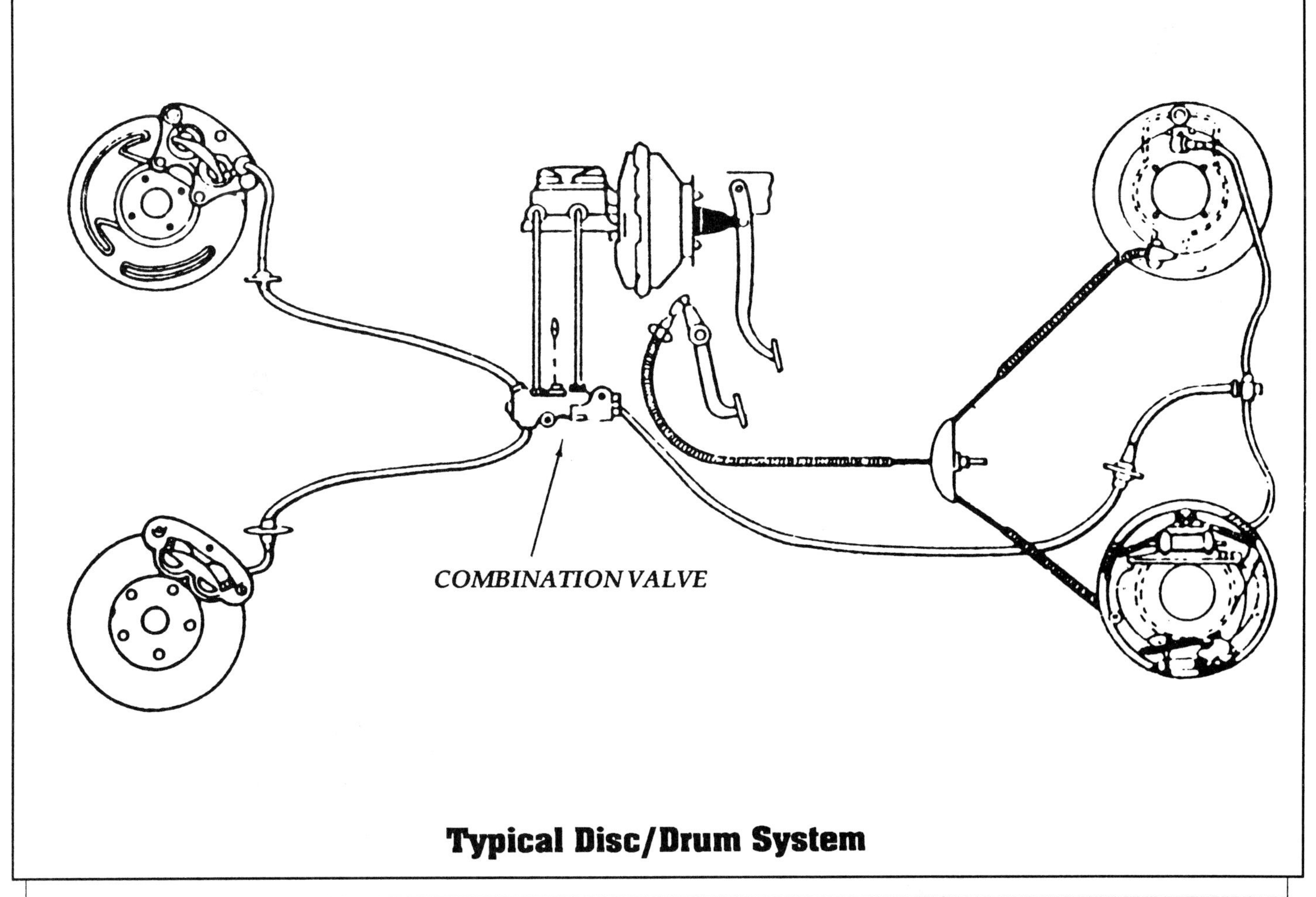

Typical Disc/Drum System

probably pitted the bores from rust. Once leakage starts, it gets bad fast. Replace or overhaul immediately. If you choose to overhaul, make sure not to hone the cylinder excessively, or the new cups may fail to seal. The manufacturer should be able to give you guidelines in this area.

2. If your system appears to be operating just fine and you add fresh fluid, be aware that this is when a borderline bad master cylinder or wheel cylinder will be first noticed. The reason is that the old fluid in the system has thickened with age, making it easier to seal. The new fresh fluid is much thinner and harder to seal. It is not uncommon to discover the need to replace a hydraulic component after new brake linings have been installed.

3. Check the emergency brake cable for rust and deterioration. Most likely, for safe operation, you will find need for repairs in this area.

4. On the '49-'51 models with the brake pedal under the floor, check the bolts holding the pedal to the master, and replace if necessary. Most likely, weather has taken its toll here. If a rusted bolt breaks during use, you could lose the brakes. This is an important safety item.

5. When performing a brake job, don't forget to change the return springs. Return springs are tempered from heat during braking, and lose their ability to return the shoes from the drums. Premature brake wear could result, if you leave old springs in.

NOTES FOR THE MODIFIED STREET BUILDER

If you are building a mild modified street machine, here are a few ways to make simple upgrades to the brakes to help you deal with the added performance.

1. On '49-'51 Fords, install the wider front shoes.

2. On cars not equipped with power brakes, add the stock power unit. Remember to put an inline check valve in the source line from the motor. This will keep sufficient vacuum in the system for one more brake actuation if the motor should die. The stock power assist unit from 1954 can be retrofitted on any of the shoebox Fords.

3. If you are installing a non-stock motor in a Ford with the master under the floor, make provisions to shield exhaust heat from the brake lines. Boiling brake fluid will definitely cause brake failure.

4. Check all of the items mentioned under Notes For The Restorer.

NOTES FOR THE HIGH PERFORMANCE BUILDER

If you are building a high performance street machine that will substantially increase the deceleration and acceleration loads, make sure all of the following items have been considered.

1. Upgrade any existing stock components as mentioned in the two previous Note sections.

2. If you are going to adapt a subframe from a newer car, make sure the components come from a car with equal or greater weight than what you are building. If you don't know exactly what you will end up with, play it safe and use the larger choice of brake components. If you intend to just adapt later disc brakes to your existing car's spindles, check the donor car to determine what the manufacturer used for spindles and bearings. If your stock components don't match favorably, you may be creating a potential stress problem. Many areas of the shoebox Ford frame and suspension systems are not stressed to deal with the loads of acceleration and deceleration as we drive today. Check with experts to ensure safety.

3. If you are radically increasing tire size on the rear axle (more than one size different than the front), remember that the brake torque requirements to stop the wheel is being altered and the brake output will have to be increased. Larger brake shoes and larger wheel cylinders are two of the possibilities. Since the design of the shoebox bodies does not allow the wheel to be moved out very much, any increase in tire width will have to be done inboard. This will require special positive offset wheels in order to keep the tires tucked neatly inside the wheel openings. (Refer to *Hot Rod Mechanix*, July/Aug 1989, page 58, for a detailed article about wheel specifications). Make sure you take careful measurements and can find wheel components that can be used, because you can be sure that this will be a problem.

One final note. Brake systems on shoebox Fords were definitely adequate for their day. The system was simple and worked well. Proper maintenance on a stock system will give you a very streetable car. If you intend to make modifications to improve the performance, however, don't be content with the brake system as it is. The brake system may only be important to you one time, but make sure it is there for you when that time comes.

Note: Special thanks to Ray Marler of Shoebox Ford Parts, 9319 S.E. 29th St., Midwest City, OK 73130; (405) 732-6027, for helping supply the technical information to ensure the accuracy of this article.

WE BUILD 'EM...

WE KNOW WHAT IT TAKES!

DO YOU HAVE A TRADITIONAL STYLE HOT ROD, CUSTOM CAR OR ANTIQUE? WORLD'S LARGEST "STRICTLY 50'S" CATALOG CONTAINS...

★ The Parts You Need ★ The Books To Show You How ★ The Tools To Help You Do It

CATALOG — $3.50

50's T-Shirt — Black w/Hot Rod & Custom Car Design, Slogan — *"We're not re-living the 50's ...we never left."* **$11.95 POST PAID** (State Size)

FREE CATALOG WITH T-SHIRT ORDER

HOT ROD & CUSTOM SUPPLY

† 1020 S.E. 12TH AVE., CAPE CORAL, FLA 33990 813-574-7744

by Tony DiCosta

PERFORMANCE FLATHEAD REBUILD

Ford flathead. Granddaddy of the modern speed equipment business. A vital link in the evolution of today's street rod. The name conjures images of the dry lakes, of smiling young men in Levis, their machines covered with the dust of Muroc or El Mirage.

The flathead carved its niche at the first dragstrips — Saugus, Santa Ana, Baylands — powering the first top eliminators.

On the street, at Bob's Big Boy, or on Whittier Boulevard, the night air carried the flathead's distinctive "rap" as the roots of modern street rodding grew.

History, you say? True, but for many of today's rodders a sense of the past, tradition, and a vehicle's ability to transcend time are all part of the allure of our hobby.

Historical criteria aside, why should you keep the flathead V8 in your shoebox? Consider the following: No engine can beat a full-dress flathead for looks; and the sound of a "built" flattie is legendary! Henry Ford's cast iron wonder, the first V8 in mass production, was and is today a viable hot rod powerplant. If your shoebox rolled off the assembly line with a V8, it had the refined version of one of automotive engineering's finest powerplants. In its final form, the 239 cubic inch Ford put out 100 horsepower and 196 ft/lb of torque. The longer-stroked Mercury, at 255 cubic inches delivered 112 horses and a whopping 200 ft/lb of torque (at a mere 2000 rpm). Both engines were conservatively rated and tuned. Both were quiet, strong, dependable, and smooth.

This engine was built by Hot Rod And Custom Supply, and was installed in a '50 Ford. Looks wild, but was totally streetable.

Within the stock powerplant is the potential for many more streetable horses and gobs of torque. Because of its long stroke in relation to its bore, the flatmotor was really a torque motor, and torque makes for a good street engine.

On the following pages, we are going to suggest how to achieve different levels of performance. All suggestions will pertain to usable, reliable, street engines. Remember, if you can't drive it, it ain't a car!

If you have a sound engine with good compression, there is a myriad assortment of speed equipment available that will pep up even a stock engine. If you just want to dress up your engine, you will be amazed at the number of goodies available for that purpose. If your engine requires machine work, most competent machine shops can handle the job. The flathead is basically a simple engine, and if you are so inclined, reassembly is well within the capabilities of most do-it-yourself type rodders.

There will be no attempt here to detail the step-by-step rebuilding process. The best book we've found on the subject is *Rebuilding The Famous Ford Flathead* by Ron Bishop. Additional information from a hot rodder's standpoint is contained in the excellent article written by Don Francisco and carried in three installments in the April, May and June 1956 issues of *Hot Rod* magazine. This article, entitled "Building a Flathead The Right Way" is available in reprint form from Peterson Publishing Company.

Before we discuss the equipment available, let's cover some basic engine combinations. The stock size Ford engine is a good place to start. Any of the performance parts we'll look at will go a long way toward increasing the performance of a fresh engine built to stock specs (3-3/16" bore by 3-3/4" stroke). There is plenty of potential here.

One step up the performance ladder is the '49-'53 Mercury engine. The Merc's 4" stroke gives it 255 cubic inches, as compared to the Ford's 239 cubic inches. The longer stroke also supplies more torque. Both engines use the 3-3/16" bore size. The Mercury crankshaft will drop right into the Ford block, and when used with Merc pistons will for all intents and purposes make a Mercury engine out of the Ford.

The last combination, and by far the most powerful we'll discuss, is made by using the 4" Merc crankshaft in combination with an overbore

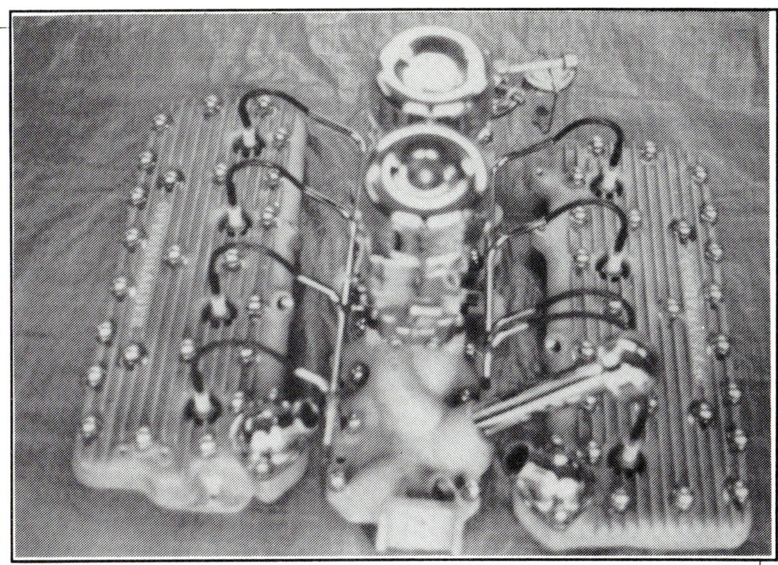

Shown are just a few of the pieces available to the flathead engine builder.

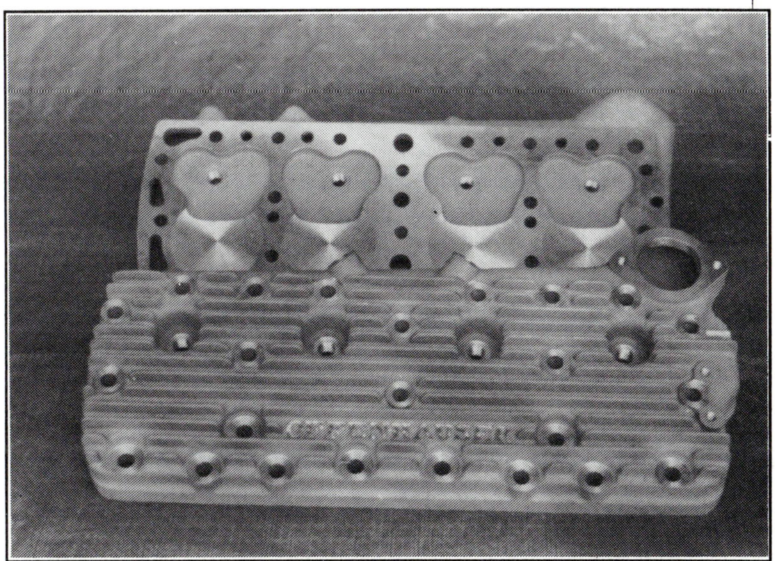

Offenhauser still produces finned aluminum heads. They are available in various compression ratios.

These are two of the best intakes for street use. Offy still offers both. The quad intake has been painted.

Commonly known as "deuces," on the left is the Ford/Holley/ Chandlers Grove 94. On right is the Stromberg 97.

There are numerous books and articles that detail the rebuilding process.

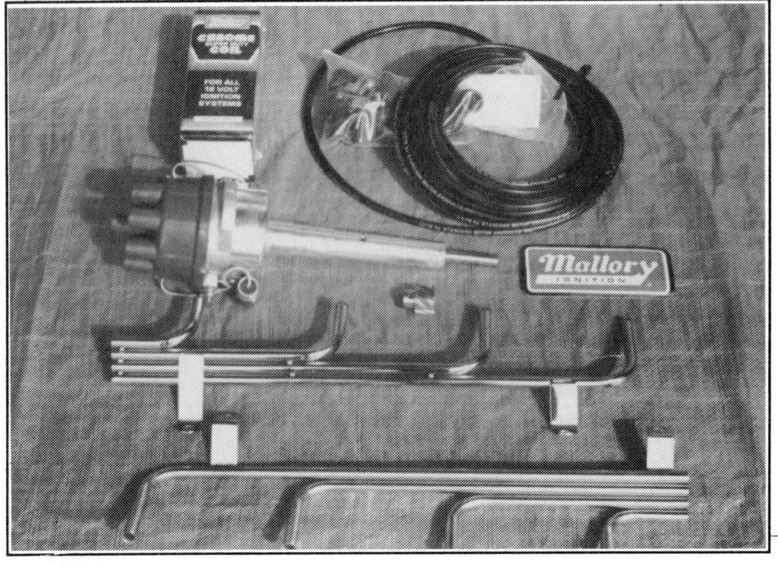

of .125. This yields 276 cubic inches, and let's remember that there is no substitute for cubic inches.

The Ford or Merc block was over-designed and is very thick in the cylinder wall area. Unless you have a very corroded engine or one that was improperly cast, most all blocks will handle the bore to 3-5/16". High performance flatheads are frequently bored to 3-3/8" with no problems. At 3-5/16" you will experience no excessive heating problems and can still "clean up" bore the engine, should the need arise.

The 276 cubic inch combination is as streetable as a 239 cubic incher, and in truth has a much easier time in the engine compartment because it is not working nearly as hard to move your relatively heavy shoebox from stoplight to stoplight. This size engine pulls like a mule going up a hill.

PORTING, POLISHING AND RELIEVING

No discussion of flathead engine rebuilding would be complete without touching on the time-honored practice of porting, polishing and relieving the block. Porting and polishing pertain to the opening up and smoothing of the intake and exhaust ports and passages. Relieving involves removing excess material from the block's deck in the area between the valves and the cylinder.

While there are many different approaches to these operations, the key to unlocking the full power potential in your flathead lies in just these areas of block prep. Because of the inherent restrictions of a valve-in-block design, anything you can do to help it breath goes a long way toward making power.

A word of caution is necessary, however. These are operations that should only be performed by someone who knows what he's doing. Other than a light smoothing of the valve pockets and intake and exhaust runners, it is best to leave the more extensive modifications to someone with experience. You can easily ruin a perfectly good block by a heavy-handed approach here.

Don't despair, however, if you don't have a legendary type in your area to perform these operations. You can build a great running (and sounding) flathead without the major block work. The parts you need to do this are only as far away as your mailbox or telephone.

You can still buy brand new Mallory dual-point distributors and coils. Also shown are a clear red plug wire set, new distributor gear and spark plug wire looms.

THE MAIL ORDER BUILD-UP

Because of the flathead's prominent role in the growth of the speed equipment business, an incredible array of high performance parts have been produced for it. A great many of these parts are still in production or are being remanufactured. There are numerous reputable mail order companies with the expertise to advise you as to the selection of components that will work together. A few companies specialize solely in flathead speed and custom accessories. These companies advertise in books and magazines. Find a company that is willing to work with you during the build-up of your engine. Some outfits will even work with your local machine shop, and can provide the expertise needed to prevent mistakes being made as the result of unfamiliarity with a high performance antique engine.

THE BASIC PLAN

Increased performance will come about as a result of improvements in three basic areas.

I - Breathing (getting air in and out), which consists of:

A. Induction (carburetion, intake manifold, etc.)

B. Valve train (camshaft, springs, valves)

C. Exhaust (headers, dual exhaust, low-restriction mufflers)

II - Compression (heads)

III - Ignition (distributor, coil, plug wires)

INDUCTION

In the induction area, you have a good selection of intakes to choose from. Two-deuce, three-deuce, and single 4-barrel intakes are currently manufactured by Offenhauser Equipment Co. and are available through their many dealers. Four-deuce intakes, high-rise intakes, and a large assortment of manifolds were produced in the "old days" by many manufacturers, and are frequently found at swap meets.

A word to the wise in this area. Many would-be street heros pick their induction system with aesthetics as the primary factor. Wrong! If you are in love with the idea of 3 or 4 deuces on a street engine, and have gobs of patience and know-how, feel free to go for it. However, almost no street flathead needs more than 2 deuces or one 4-barrel carb. Two deuces will give you the multiple carburetion look without the hassles of a more

Jahns makes flathead pistons for any bore and stroke combination you can build. Grant rings have modern 3-piece oil ring.

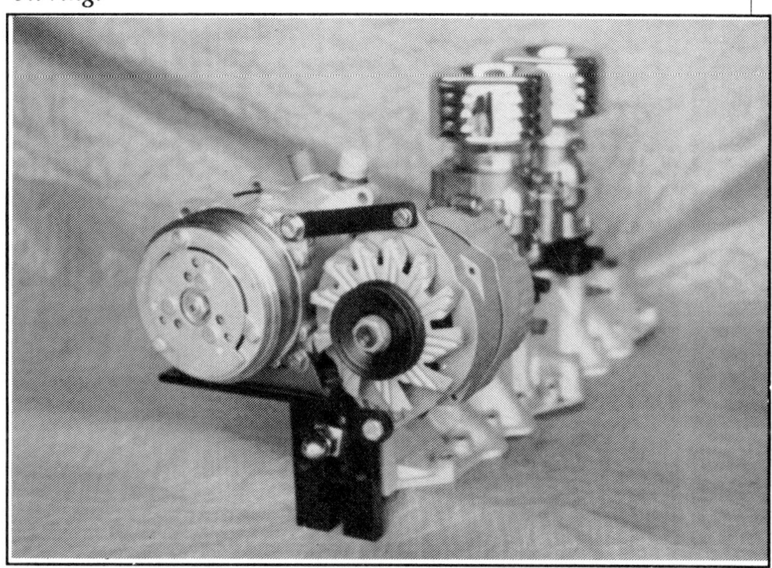

You can have bolt-on air conditioning too. The bracket shown mounts both A/C compressor and GM alternator. Available from Hot Rod And Custom Supply.

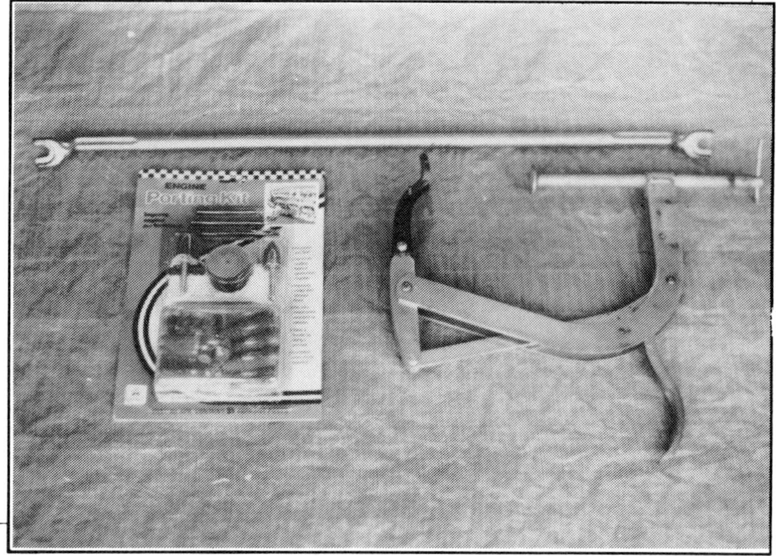

Engine rebuilding aids: Valve removal bar, engine porting kit, and valve assembly tool.

complicated set-up, and will run much better around town. A 4-barrel intake with a small quad will improve performance and is low maintenance to boot.

As for carburetors, Stromberg 97's may have a legendary appeal, but they are famous leakers and can be a fire hazard when they malfunction. Any of the later 3-bolt carbs, commonly referred to as 94's (this includes Ford, Chandlers Grove, and Holley carbs) are easier to find in good condition and will work just as well if not better than Strombergs. A good motorcycle type fuel filter is a must, as all these carbs are extremely dirt-sensitive.

A stock fuel pump in good condition will work just fine. Replace the fuel pump pushrod, however. If you use an electric pump, you must use an adjustable fuel pressure regulator.

VALVE TRAIN

As with any engine, the valve train is really the heart. A good camshaft will really wake up a flathead. A discussion of camshafts alone would fill a small book, and there were a lot of really great cams ground for the flathead. Modern camshaft design was pioneered by flathead racers. Such legendary "sticks" as the Potvin Eliminator and the Iskenderian 404 were considered the zenith of camshaft manufacturing. Many excellent cams are still available from a variety of manufacturers. However, keep in mind two things: 1. Used swap meet cams are not a good investment. Most of us have no way of checking the wear on these items, and rarely do swap meet cams come with a spec sheet. 2. Don't buy a cam for its name or image. The two legendary cams mentioned previously, for example, were never intended for use in a heavy street car.

The selection of a cam should be made after consulting with a reputable parts purveyor and taking a number of factors into consideration. Also, it's best to choose a cam at the beginning of a build-up.

Never use stock valve springs with any performance camshaft. Get the cam grinder's recommendations for the proper springs. A slightly larger 1.60" valve (stock is 1.50") that is undercut and backcut will fit a stock reground valve seat and offer increased air/fuel flow without sacrificing velocity. Modified Chevrolet valves of this type are available from numerous sources, as are bronze-bushed valve guides.

EXHAUST

A flatmotor's exhaust system is important for two reasons. Increased breathing is a key to power, but cultivating the proper flathead sound — an engine that talks to you — is one of the great pleasures of owning this legendary engine. It's because a flathead's exhaust is routed through the block that the exhaust note is deeper than other engines. A flattie with "the sound" will stop traffic at the busiest rod run. The proper exhaust system will capitalize on this potential.

A good selection of headers is available, ranging from well-made steel tube pieces to the high quality Fenton cast iron headers. Exhaust baffles are a good investment. Several types are available. Pick the smallest, as these will help direct the exhaust charge without becoming restrictions of their own. It has been shown that baffles (or dividers) can help to reduce the hot spot that forms between the two middle exhaust valves.

Dual exhausts are a must for both sound and power. Most street flatheads really seem to like 1-3/4" tailpipes. Since mufflers are such an important part of the flathead sound, we are going to make a few specific recommendations. Currently available Smithy's mufflers (a repro of a great '50s item) will give the deepest, mellowest sound. Brockman Mellow-Tone steel packed mufflers are a bit louder, but also have a beautiful deep tone, unlike today's current crop of glass packs and turbo mufflers.

A set of chrome echo cans (3-1/2" by 12" is best for shoeboxes) will deepen the tone even more and help produce the "rap" that is music to a flathead owner's ears.

COMPRESSION

At a compression ratio of 6.8:1, stock flatheads could use a little improvement. Even with today's gas, you can get away with 7.5:1 or 8.0:1 without a problem. This is where aluminum heads come in. In addition to their great looks, these finned heads offer the added advantage of greater cooling capacity, better heat dissipation and lighter weight. Both Offenhauser and Edelbrock still offer these items (Offy in a variety of compression ratios).

In regard to the purchase of a set of heads, here is some advice that many have paid dearly for. When you are budgeting money to build an engine, allow enough to buy a new set of heads. There are numerous reasons for this. Suffice it to say that unless you carry around a machinist's straight-edge and have the benefit of years of experience buying used flathead parts, a junk set of used heads will look just like a good set of used heads. If you're on a budget, scrimp or sacrifice in another area until you've covered the cost of a set of heads. The money saved by buying a used set of heads probably will not be worth the aggravation.

IGNITION

The builder of a '49 to '53 flattie has some excellent options in the ignition department. These options, however, shouldn't include the use of an unmodified stock distributor. From a performance standpoint, this was one of Henry's poorest pieces. They used a

this was one of Henry's poorest pieces. They used a complicated system within the carb (some even used an air valve) to operate a cantankerous vacuum canister. Using anything but a stock carburetor will mess with the vacuum signal and the stock single-point distributor will not advance properly.

Because of a flathead's naturally shrouded spark plugs (they're practically in a hole), they need a strong spark. The longer dwell achieved with a set of dual points is a great remedy for this problem. Many good aftermarket distributors were made, and these frequently show up at swap meets. Distributors found at swaps can be field inspected on the spot for wear, and many competent auto shops can tune them on modern testing equipment. Switching to 12 volts is not a bad idea, either.

One of the oldest names in the hobby, Mallory, offers an excellent centrifugally advanced dual-point distributor that is reliable and efficient. Replacement parts are readily available.

If you're a thoroughly modern kind of person, breakerless ignitions are also available. The same array of high performance ignition components that are available to a modern engine builder is available to the flathead owner. You can have high output coils, electronic ignition controls and multiple sparks just like the "big boys."

For an old timey look, clear red spark plug wire sets are available, as are polished stainless steel wire looms, to keep the ignition wires off hot engine surfaces.

If you want to drive in air conditioned comfort, brackets are available to mount low-drag compressors and alternators to the engine.

DRESS-UP GOODIES

No engine dresses up quite like a flathead. If you are looking for a "drop dead" beautiful engine, the list of shiny pieces for a flathead is almost endless. From the time-honored chrome acorn nut covers to finned coil covers, from chrome radiator pipe sets to chrome oil fill tubes, this is an area you can really go hog wild in.

Maybe you're undecided as to whether or not you'll keep that historic flathead, or do a "chain-fall" tune up. If that's the case, take in the photos on these pages before you decide.

Feast your eyes on these goodies. All items shown are currently available. Imagine how these would look on a real engine.

This block has been ported, polished, relieved and ground. It's so beautiful in this phase that it's almost a shame to put the rest of it together.

by Pat Ganahl

ENGINE SWAPS

courtesy of Butch's Rod Shop

Back in the '50s, all sorts of companies — Hurst, Almquist, Newhouse, Honest Charlie, and so on — made engine mounts, tranny adapters, dropped spindles, lowering blocks, dropped tie rods, and plenty of other custom parts to bolt Cads, Oldses, Chevys and who knows what else into shoebox Fords and bathtub Mercs. But all of that stuff has been made of unobtainium for years.

When it came time for street rod manufacturers to reinvent bolt-in engine swap kits, one of the first to respond was Butch Bunn of Butch's Rod Shop in Dayton Ohio. Being a devotee (and owner) of '49-'57 Fords and Mercs, he decided to tackle the project of swapping both small block Ford (289-302) and small block Chevy engines and transmissions into each of these models.

Although the swap kits and their installation are surprisingly simple, designing them was a chore. The problem is that Ford and Merc chassis changed several times between '49 and '57. Plus, '49-'53 flathead cars had rear-sump oil pans, while '54-'57 Y-blocks had front-sumps. To cover all applications required Butch to design eight separate kits.

Our photos primarily illustrate a small block Ford engine installed in a '49 Ford and a '54 Merc, but here is a quick rundown of '49-'57 FoMoCo differences:

• '49-'51 Ford — Large forward-mounted front crossmember; integral trans crossmember; independent front suspension with kingpin spindles. All '49-'57 Fords and Mercs use a 3-link, idler arm

Whether you want to swap a Chevy or Ford small block into your '49-'57 Ford or Merc, Butch has you covered with bolt-in kits.

The swap kits are simple and inexpensive. This is the kit for swapping a Ford small block engine into a '49-'51 Ford.

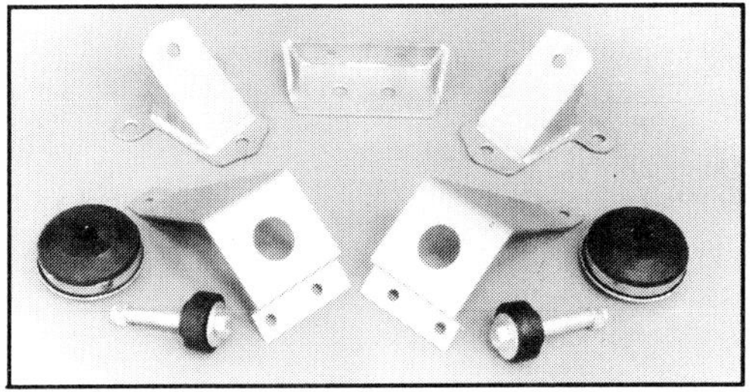

Here is the kit for installing a small block Chevy engine in a '49-'51 Ford chassis.

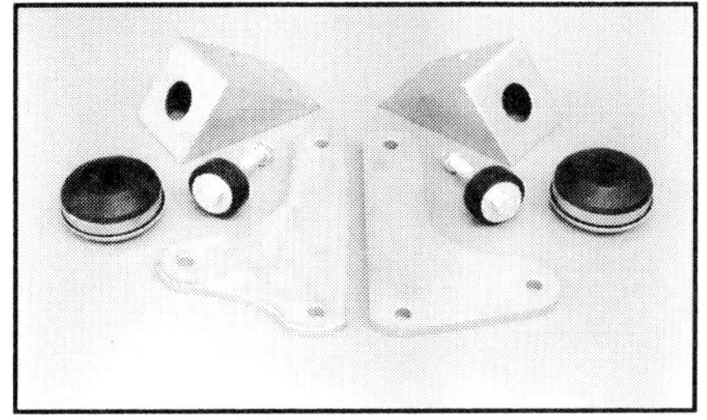

If you want to install a small block Ford powerplant into a '52-'53 Ford or Merc, this is the kit you need from Butch's Rod Shop.

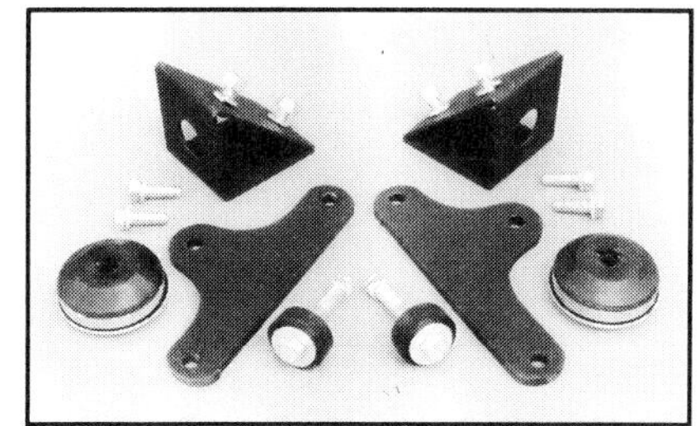

Stuffing a small block Ford engine into the chassis of a '54-'57 Ford or Merc requires the use of this kit.

tie rod setup, and the center link causes oil pan clearance problems in most cases. In this case, the Ford engine requires a Bronco rear-sump pan, while the Chevy requires that the pitman arm and idler arm be heated and carefully bent to lower the center tie rod link 3/4" to 1". A rare V8 Nova pan would solve the problem, but they are very hard to find.

• '49-'51 Merc — Similar design, but suspension components (kingpins, spindles, etc.) are larger, and the front crossmember and steering linkage is farther forward; drop-out trans crossmember. The Chevy fits easily in these cars, as is. Butch is R&D-ing the Ford kit, which is a more difficult swap, and will probably require weld-in mounts.

• '52-'53 Ford and Merc — Both use same kingpin front suspension. Ford engine compartment lengthened like '49-'51 Merc. Drop-out trans crossmember. These cars accept the front-sump Ford pan. For clearance with the Chevy, the pitman and idler arms can be retapered to mount the center link below, rather than above them.

• '54-'57 Ford and Merc — New balljoint front suspension. Steering remains basically the same. Drop-out trans crossmember. Tubular engine mount crossmember. Front-sump Ford engine drops in easily. Chevy requires steering center link to be lowered 1' to 1-1/2", stock mounts to be cut off crossmember, and new kit mounts to be welded in place.

The '49-'51 Ford swaps require the transmission hump in the floor to be raised about 1-1/2". All swaps necessitate reworking or possibly replacing the bolt-in cover in the floor for trans clearance.

Each of Butch's kits uses the common early Ford biscuit-type rubber mount. Not only does it make engine installation easy, but the through-bolt ensures that the mount won't pull apart, and it limits engine torque twist.

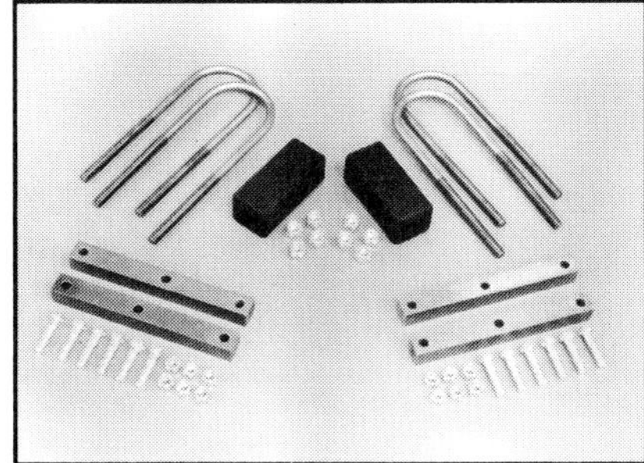

Butch also offers lowering block kits for the rear axle in various sizes. The 1" bars below them will lower the front of '49-'56 models about three inches when installed between the lower A-arm and the spring retaining plate (after rivets are drilled out).

This is the engine compartment in Butch's clean '54 Merc. The '54-'57 Fords and Mercs are the same, with a tubular engine mount crossmember and drop-out trans mount. These were designed for front-sump engines.

The small block Ford fits in these cars as if it were intended to. The kit mounts bolt right into the crossmember. The Chevy is more difficult, requiring stock mounts to be cut from the crossmember, and new kit mounts welded to it.

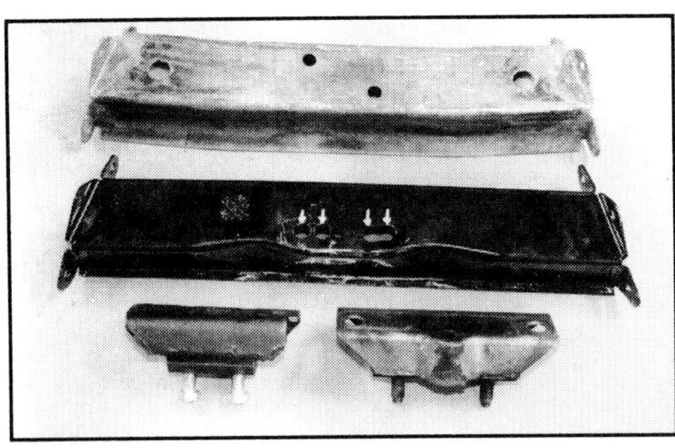

The '49-up Mercs and '52-up Fords use a bolt-in trans mount member, which is redrilled in most of Butch's kits to accept a Ford or GM stock trans mount.

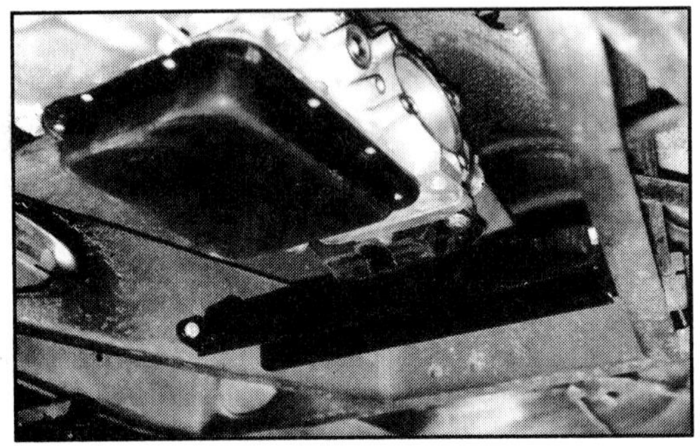

Trans installation couldn't be simpler.

Close-up of front engine mount, viewed from front of car.

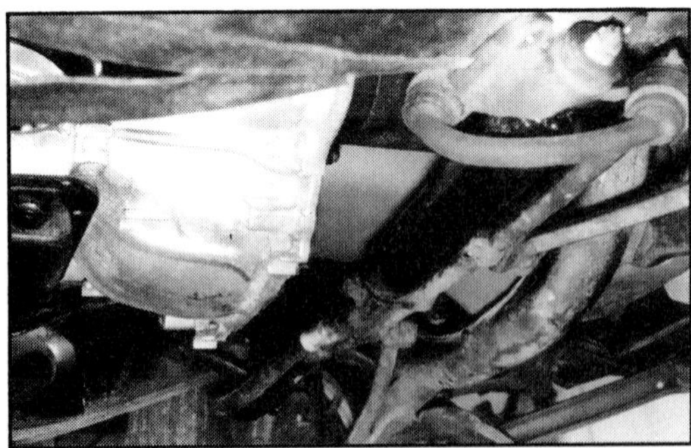

Front-sump Ford pan affords plenty of steering clearance, even with rare power steering option. The Chevy requires lowering of the center link about 1-1/2"; power steering won't fit.

The small Ford looks nearly dwarfed in the big engine compartment. The same kits should swap the taller-deck 351 Windsor into these cars, and very likely the 351 Cleveland.

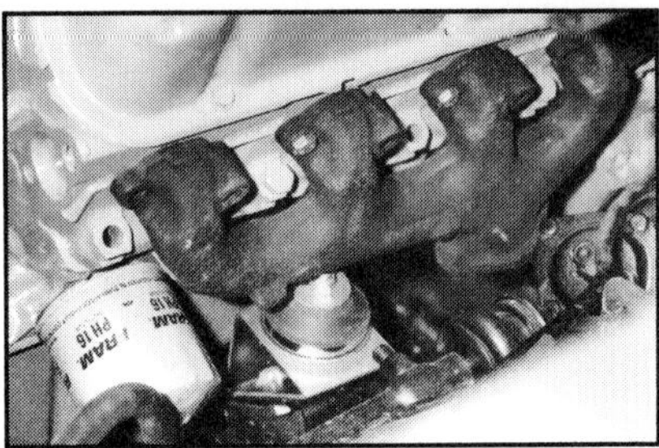

All kits allow use of stock exhaust manifolds. Instruction sheets recommend specific types for certain applications.

Overall underside view shows perfect fit of front-sump Ford and layout of suspension common to '54-'57 Fords and Mercs. Note that longer tie rods attach closer together on the center link, compared to '49-'53s. This eliminates bump-steer. Fat Man's tie rod kit for '49-'53 does the same.

One way to lower the front of '49-'56s is to install these 1" spacer bars between the A-arm and the spring plate. This lowers the car about 3" at the wheel, yet retains full spring length. Note ball joint spindle.

Our second illustration example is a clean '50 Ford 2-door.

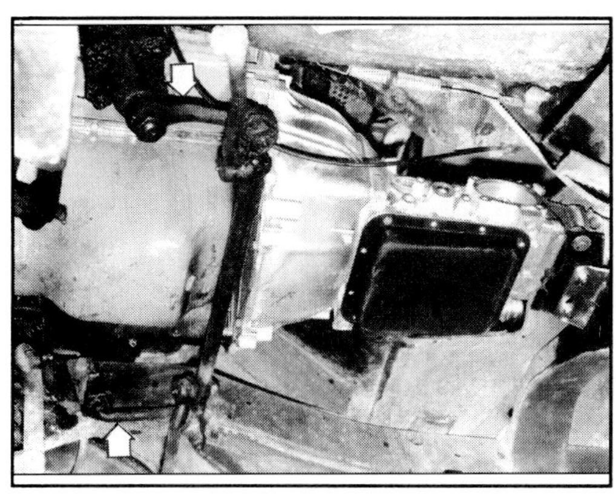

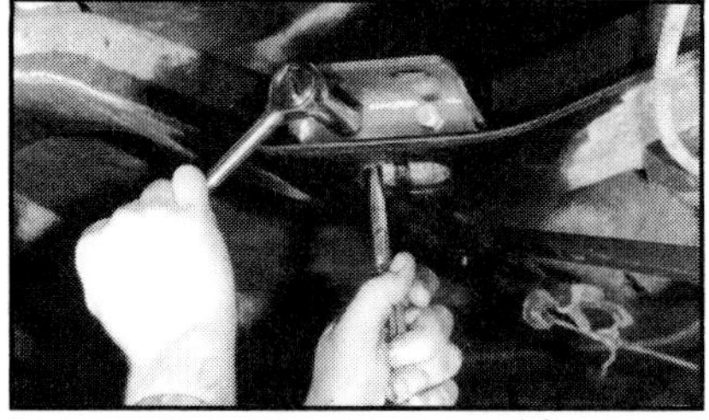

The '49-'51 Ford (only) has a permanent trans crossmember. First step in the swap is to locate the center, drill two holes, and bolt the new trans mount in place.

To fit the Ford into '49-'51 Fords requires swapping the front-sump pan and pickup for a rear-sump Bronco or '78-up Mustang unit. The Bronco, with a single drain plug, is preferable.

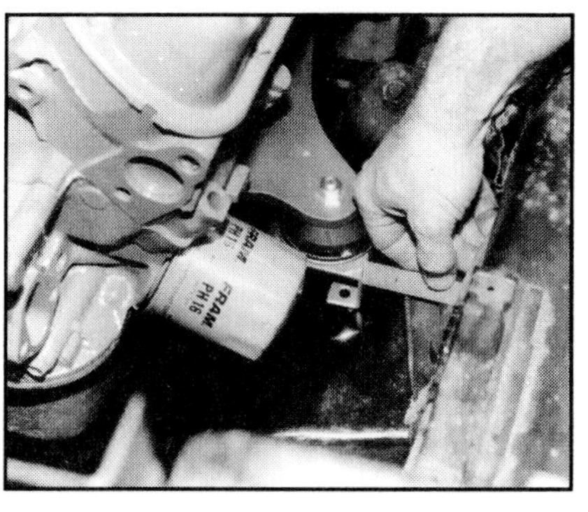

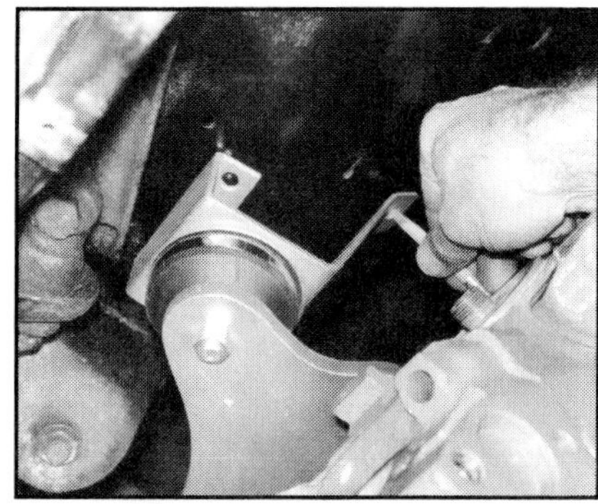

With the complete mounts bolted to the engine, slip the engine/trans in place and loosely install the trans mount bolts. Next, drop the front mounts onto the crossmember, measure distance to the frame left and right to center the engine, then mark bolt holes on the crossmember for drilling. Easy.

Remove the engine, drill the crossmember, and bolt the mounts in place. You can see how the tie rod center link gets in the way.

SOURCES

Butch's Rod Shop
2853 Northlawn Ave.
Dayton, OH 45439
(513) 298-2665

Fat Man Fabrications
8621 C. Fairview Rd.
Hwy. 18
Charlotte, NC 28227

ECi
P.O. Box 2361
Vernon, CT 06066
(203) 872-7046

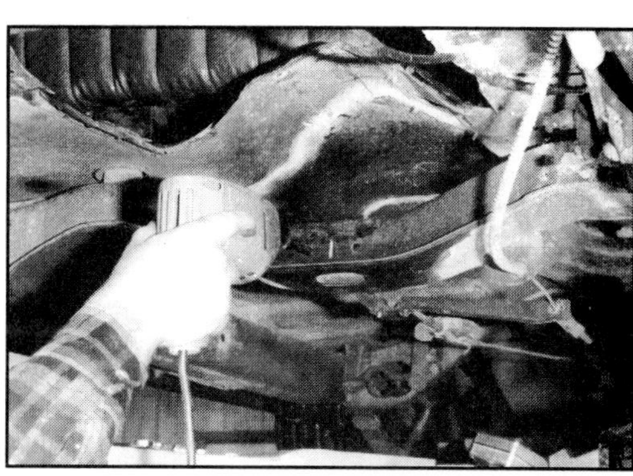

Underside view of the '50 shows how the Bronco pan neatly clears the steering. With the Chevy, the pitman and idler arms must be heated and bent down to drop the center link about an inch.

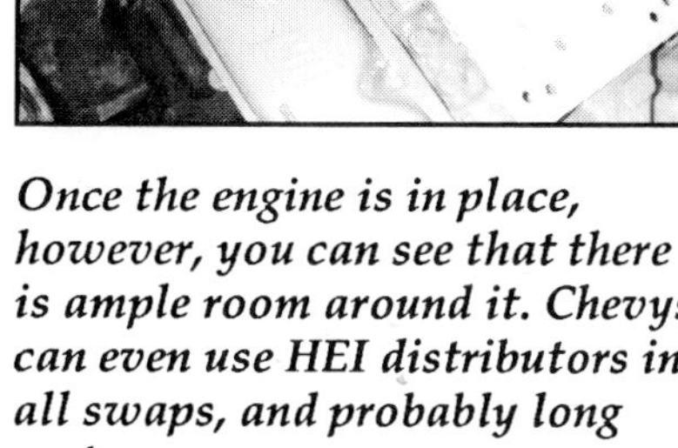

Once the engine is in place, however, you can see that there is ample room around it. Chevys can even use HEI distributors in all swaps, and probably long water pumps.

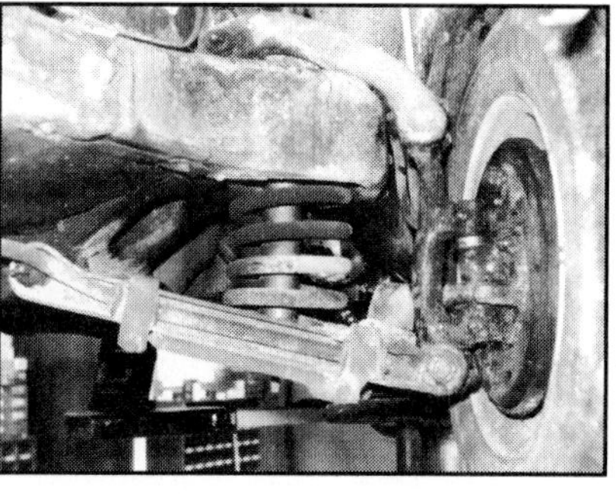

This '50 (which happens to have a front sway bar attached to the lower A-arms) shows the spindle upright and kingpin arrangement used on '49-'53 Fords and Mercs.

In a wise move, Butch's Rod Shop, Fat Man Fabrications, and Engineered Components, Inc. (ECi) have decided to pool their individual talents to make compatible parts and avoid competitive duplication. ECi specializes in brake systems, and currently offers an '82-up Camaro/Malibu front disc brake swap kit for '49-'51 Fords, with others to follow. Fat Man is the suspension specialist, and has been trying to work the bugs out of these Fords and Mercs. To solve their tendency to toe-in with chassis rise, he offers a new, correct-geometry tie rod set for '49-'53 Fords, plus new idler arms. He can also provide dropped center links for pan clearance. But the biggest news is that he has introduced 2-1/2" dropped spindles (actually dropped uprights), made from seamless .250-wall tubing with machined bosses, for '49-'53 Fords and Mercs.

Finally, Fat Man is currently considering a kit to adapt a GM power rack and pinion from FWD Somerset Regals and Olds Calais to the Ford and Merc chassis, which would solve clearance and bump-steer problems while affording new power steering. While stock '49-'57 FoMoCo boxes don't have inherent problems (and can be rebuilt), a '69-'79 Ford half-ton or later truck (F-100, 150, 250, 350) or '69-'78 Ford full-size, Torino, T-Bird, Mercury, or Lincoln power steering box will bolt in.

As for rearends, all of these cars use dual parallel leaf springs with tube shocks, so most anything can be easily adapted. Butch reports that a '71-'77 V8 Maverick 8-inch rear ('74-'77 are 5-lug) is the same width as stock (57" flange to flange) and bolts to the springs. The '65-'69 Mustang rear is the same, but they're rare and expensive. Of course, '57 Ford wagons introduced the 9-inch, and '57-'59 Ford rears are bolt-ins, but they're not easy to find either.

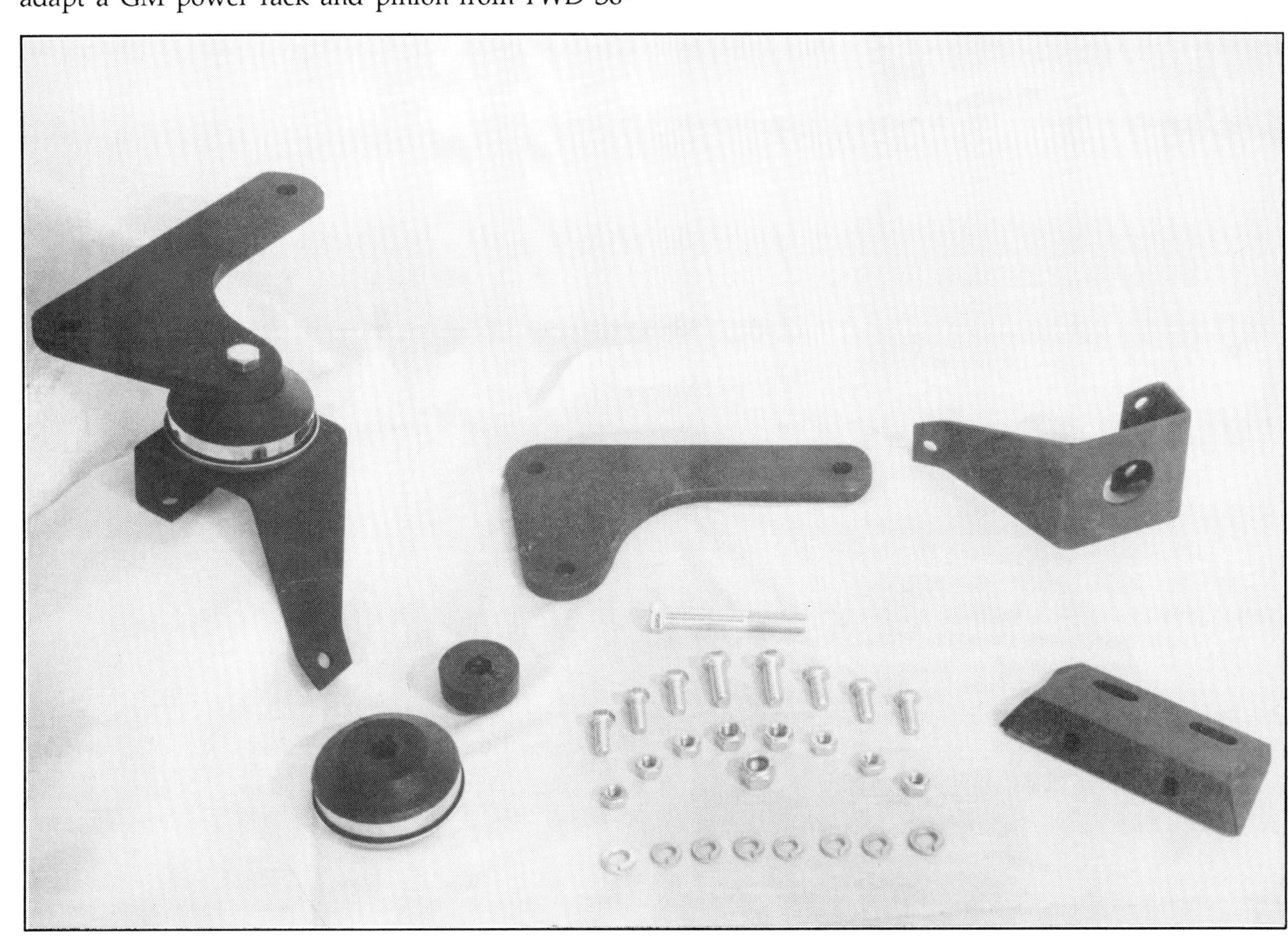

This complete bolt-in kit is for installing a 302 Ford V8 engine in a '49-'51 Ford. Also included is the transmission mount kit.

by Dave McNurlen

INSTALLING A STOCK FORD 5.0 TPI HARNESS

Let's say that somebody has a crashed '86 Mustang with a multiport fuel injected 302, and you're drooling to put the engine in your hot rod. You know about the aftermarket harness kits available, but you've got a complete car, with complete wiring and vacuum harnesses, and you'd like to use those if you could.

Can you? The answer is a resounding maybe. The factory harness is considerably more complex than a Street And Performance, or Hypertech system, with a lot more components. The Ford system must be used almost in its entirety, since the system won't go into closed loop if some of it is missing. The engine must be left stock for the same reason. Changing the cam is a no-no. If you're willing to put up with these restrictions, then no problem.

An '86 Mustang will be used as an example, since there are a lot of them out there. Cougars and Thunderbirds used the same set-up, with minor wiring changes, and this entire system stayed very similar until '90, when Ford started using a mass-airflow type system on the engines.

Ford uses two oxygen sensors on this engine, mounted at the top of the separate catalytic converters on each side of the engine. You'll have to use both sensors. If the converters aren't used, mount the sensors no farther than 18 inches away from the closest exhaust port. These are heated oxygen sensors, that go into operation much faster than conventional non-heated sensors.

An important electrical circuit component is the inertia switch. This is in-line to the fuel pump relay,

Here's the object of all this: an '86 Mustang fuel-injected 302. The aluminum valve covers are stock, and the rubber cover for the distributor cap is interesting. The box to the right of the battery covers the coil, and the starter relay.

The engine compartment harness is visible here. The plastic loom runs from one side of the firewall to the other. It has the oval-shaped grommet on the passenger side, with a 7-wire, and 4-wire connector by the windshield wiper motor on the driver's side. The two toilet-seat looking connectors behind the word "Ford" plug into the actual engine harness. The round cap behind the passenger side strut tower is the EGR vent solenoid.

has a reset button, and cuts off power to the fuel pump in the event of a crash. It's located in the trunk on Mustangs, either by the left trunk hinge, or by the left taillight. To work properly, it should be mounted in the same position as it came off the donor car.

The fuel system must have a return hose to the gas tank, and a high pressure fuel pump. This doesn't have to be the stock Ford pump, as long as the pump used can provide at least 60 lbs. of pressure. If an in-tank pump isn't used, mount the external high pressure pump no farther than 18 inches away from the tank, and as low as possible.

The rectangular thing on the bracket at left is the MAP sensor. Besides an electrical plug, it also has a vacuum connection. The main vacuum manifold is here, with all ports clearly marked. The wiper motor is to the right.

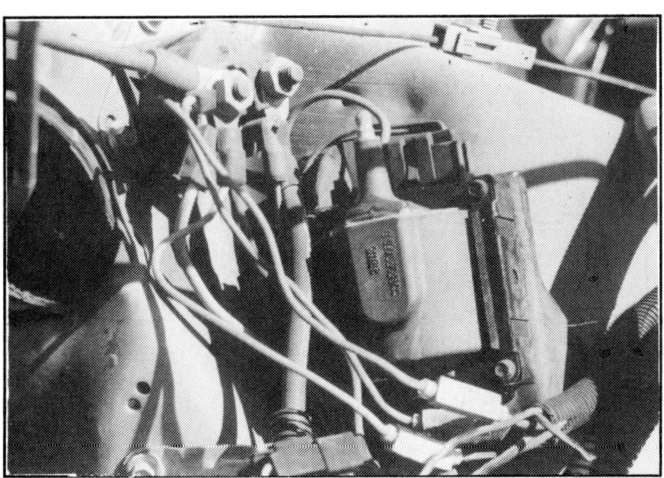

Working around to the front of the driver's side strut tower, we find the starter relay and the coil. The fusible links are all blue, with the actual wire starting after the rectangular block.

Just below the wiper motor, are the 4-pin, and 7-pin connectors that go to the column and fuse block. If you don't use the Mustang ignition switch, a suggestion would be to use these connectors to hook up to your ignition switch and fuse block, and leave the engine harness intact. The wire colors are noted on the schematic.

Visible here are the main engine wire connectors. The small can is the fuel pressure regulator, with its vacuum hose connected. The larger can is the EGR valve, with its vacuum hose unhooked. These cars have a vacuum schematic sticker under the hood, and it can come in very handy when hooking things up. The plastic part on this end of the EGR valve is the EGR valve position sensor. Also visible is the fuel pressure test port just to the left of the fuel pressure regulator.

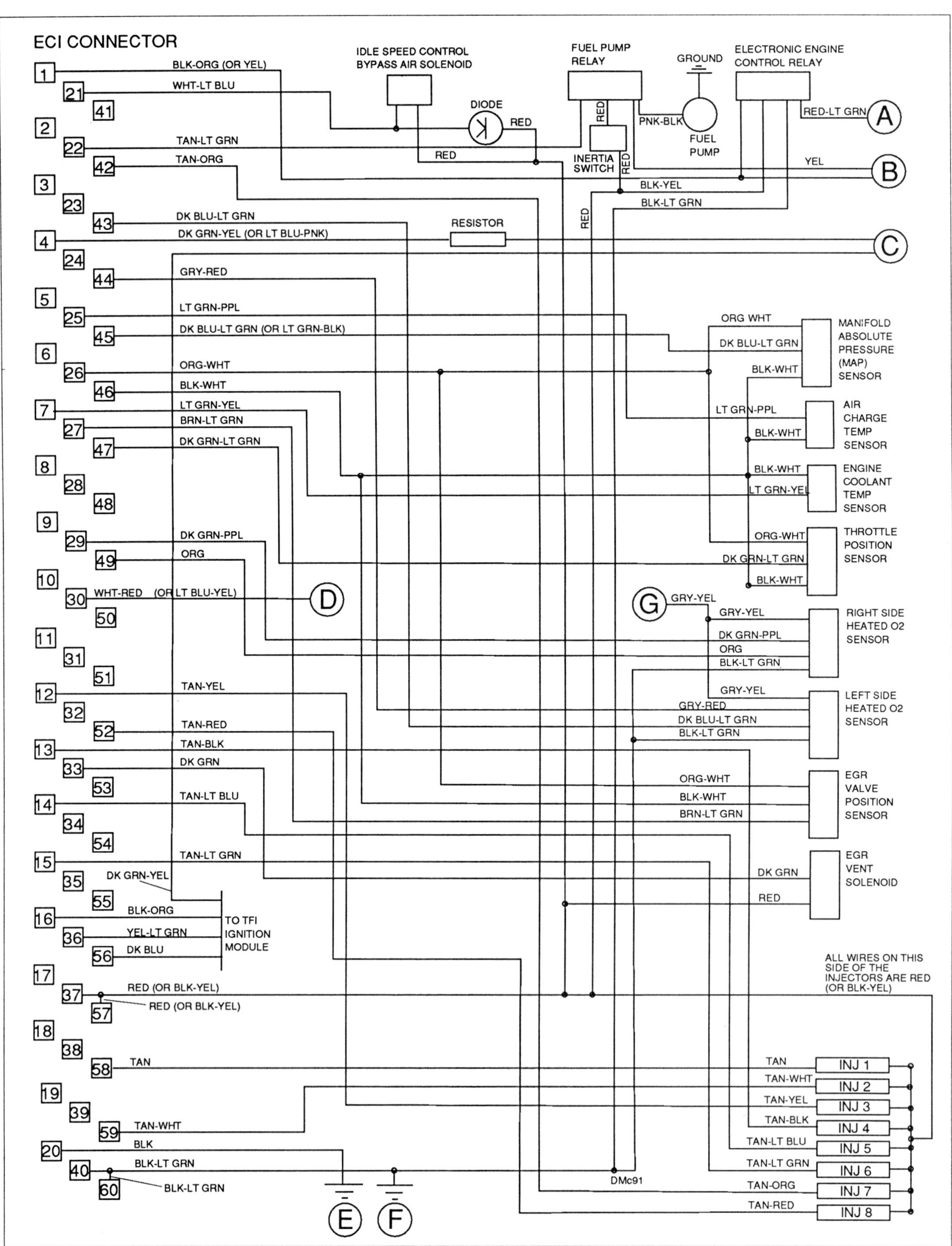

ECI CONNECTOR
IDLE SPEED CONTROL BYPASS AIR SOLENOID
FUEL PUMP RELAY
GROUND
ELECTRONIC ENGINE CONTROL RELAY
DIODE
FUEL PUMP
INERTIA SWITCH
RESISTOR
MANIFOLD ABSOLUTE PRESSURE (MAP) SENSOR
AIR CHARGE TEMP SENSOR
ENGINE COOLANT TEMP SENSOR
THROTTLE POSITION SENSOR
RIGHT SIDE HEATED O2 SENSOR
LEFT SIDE HEATED O2 SENSOR
EGR VALVE POSITION SENSOR
EGR VENT SOLENOID
TO TFI IGNITION MODULE
ALL WIRES ON THIS SIDE OF THE INJECTORS ARE RED (OR BLK-YEL)
INJ 1
INJ 2
INJ 3
INJ 4
INJ 5
INJ 6
INJ 7
INJ 8
BLK-ORG (OR YEL)
WHT-LT BLU
TAN-LT GRN
TAN-ORG
DK BLU-LT GRN
DK GRN-YEL (OR LT BLU-PNK)
GRY-RED
LT GRN-PPL
DK BLU-LT GRN (OR LT GRN-BLK)
ORG-WHT
BLK-WHT
LT GRN-YEL
BRN-LT GRN
DK GRN-LT GRN
DK GRN-PPL
ORG
WHT-RED (OR LT BLU-YEL)
TAN-YEL
TAN-RED
TAN-BLK
DK GRN
TAN-LT BLU
DK GRN-YEL
BLK-ORG
YEL-LT GRN
DK BLU
RED (OR BLK-YEL)
TAN
TAN-WHT
BLK
BLK-LT GRN
RED
PNK-BLK
RED-LT GRN
YEL
BLK-YEL
GRY-YEL
DMc91
A
B
C
D
E
F
G

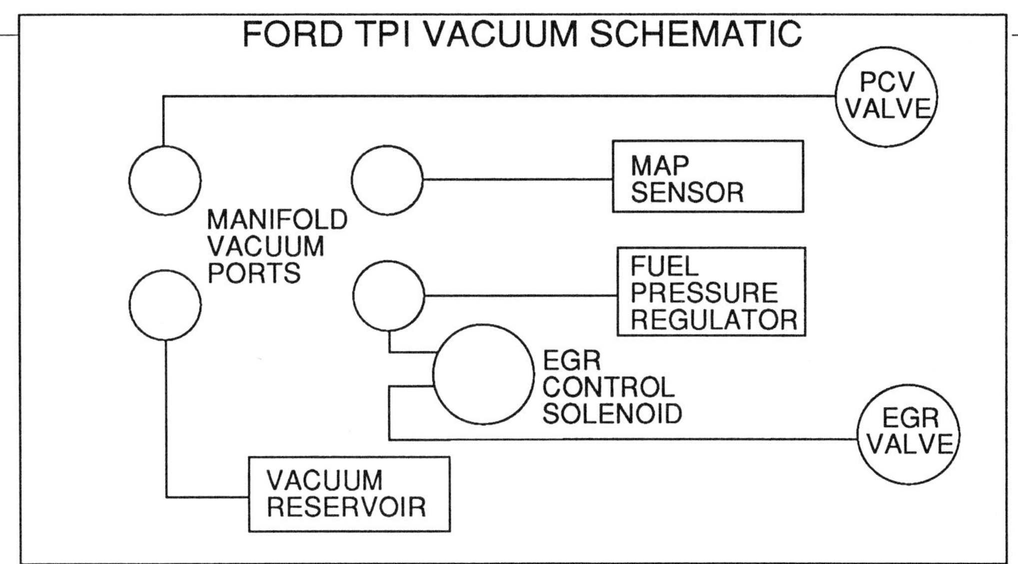

HARNESS CONNECTION LOCATIONS

A
Red wire with light green stripe; power to ECC relay. Connect to fuse-protected ignition-on; connection made at 7-pin connector on driver's side of firewall near wiper motor. On a Mustang, this connection is made at two different ignition switch posts: I1, and I2. I1 supplies power to the circuit in the run position, and I2 supplies power in the start position. I1 is red-light green stripe at the switch, and I2 is brown-pink stripe at the switch. Test the switch you're using to see if this type of connection is possible. Don't connect the I2 wire and the "D" wire together, or the starter will be engaged all the time, from feedback through the circuit.

B
Yellow wire to fuel pump relay. Black with orange stripe (or yellow) wire to #1 pin of ECA. Both blue fusible link wire connections at starter solenoid; 12 volts all the time, not switch-controlled.

C
Dark green wire with yellow stripe (or light blue with pink stripe) from pin #4 of ECA. Dark green wire with yellow stripe that goes to TFI ignition module on some cars: to negative (-) connector at ignition coil. Hook positive (+) side of coil to fused ignition-on. Power should be supplied by connection "A".

D
Light blue wire with yellow stripe, originating at pin #30 of ECA, passing through a diode, and coming out at 7-pin connector on driver's side by wiper motor, as yellow wire with red stripe. Connect to fuse-protected hot-in-start only.

E
Black wire from pin #20 of ECA: case ground for ECA. Eyelet connector located in vicinity of ECA case. This wire is visible in the photo of the ECA connector, although the eyelet is broken off.

F
Black wire with light green stripe: main ground for system located at battery. Connect to negative post of battery.

G
Gray wire with yellow stripe, from both oxygen sensors. Connect to fused ignition-on.

Stock components not used, and their wire colors for ID:
- *Thermactor air diverter solenoid: light green with black stripe/red.*

- *Thermactor air bypass solenoid: white with red stripe/red.*

- *A/C wide open throttle relay: red/orange with light blue stripe.*

Wire ID:
blk = black, blu = blue, brn = brown, dk = dark, gry = gray, grn =green, lt = light, org = orange, pnk = pink, ppl = purple, red = red, tan = tan, wht = white, yel = yellow.
First color = main wire color. Second color = color of stripe on wire. Example: Blk-yel = black wire with yellow stripe.

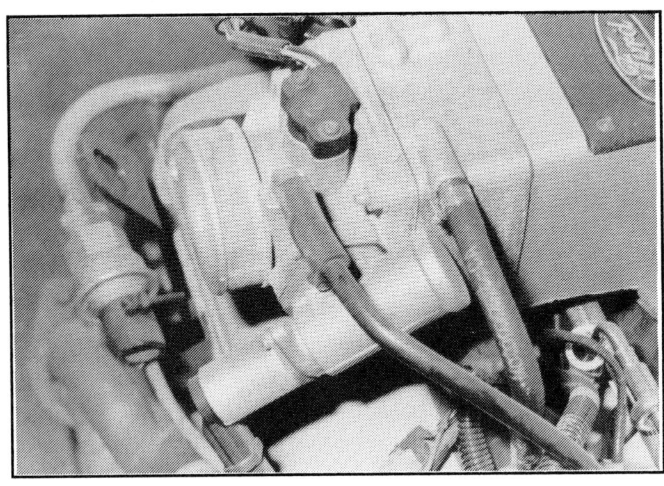

On top of the throttle body, held on by two screws, is the throttle position sensor (TPS). The long round thing on this side of the air intake is the idle speed bypass air solenoid. It lets air around the throttle butterfly to maintain idle speed.

This will most likely be what your main ground connector will look like (connection F on the schematic). This connector can corrode internally, and should be replaced with a crimped and soldered connector.

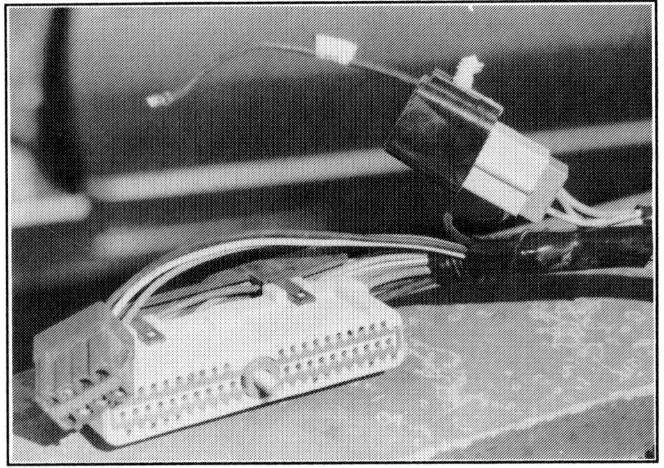

This is the ECA connector. The ECA mounts on the inside passenger side front kick panel. The single dark wire is connector E on the schematic.

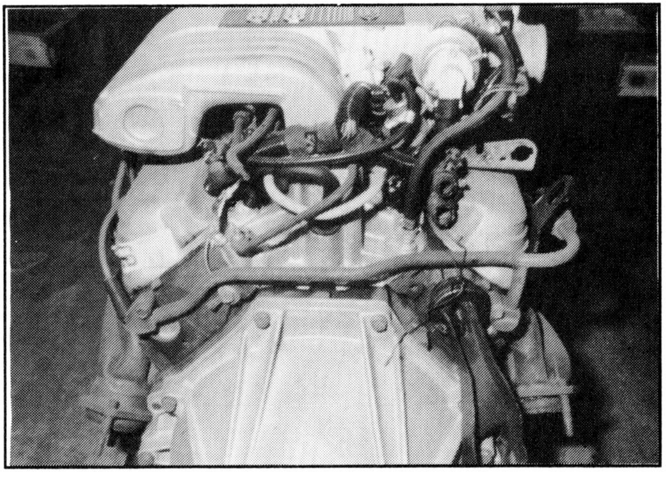

Here, the air tube on the back is visible. This pipe can be removed, and the ports in the heads blocked off. None of the air pump system is used.

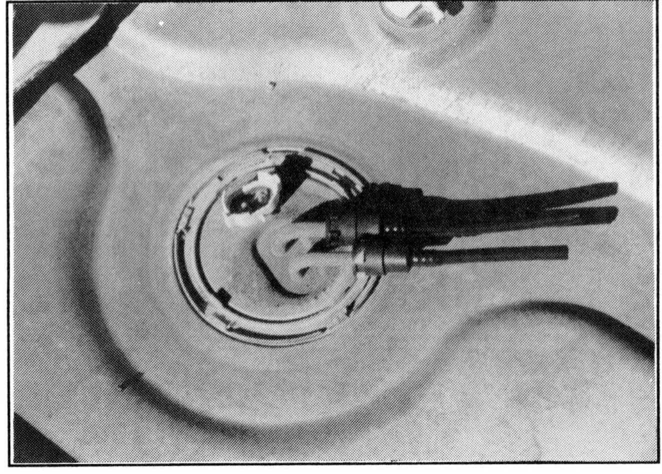

This tank unit shows the fuel lines and fittings Ford uses. Fittings similar to these are used from front to back. The lines are a flexible plastic-like material, and the connections just push together; held together with an interlocking spring system, and sealed with an O-ring. Special tools are available from most tool companies to get these connections apart. Most of the hard lines can easily be double flared, to use conventional fuel lines and a 150 psi rated high-pressure fuel hose.

TEX SMITH'S HOT ROD
MECHANIX

by Ray Marler
courtesy of 49-50-51 Ford Owners Newsletter

FLATHEAD

COOLING SYSTEM

I bring to you an eternal truth: Some flatheads are going to run "hot" no matter what you do.

But just because some run hotter than you might believe normal is no reason to throw up your hands and quit. In most cases, I've found there's a basic reason a flathead overheats, and what this article is all about is how to find that reason and remove it. Not all flatheads run hot, of course; some owners never experience problems. Yet we know that hundreds of owners face some kind of cooling system problem today.

THE BASICS

Some basics: An engine is a producer of heat. That's where it gets its energy. As the combustion chambers explode with the fuel/air mixture, heat is produced. Friction adds to this heat. The removal of that necessary heat is essential, or the engine will suffer meltdown. To prevent this, internal combustion engines have water jackets or passages surrounding the cylinders. Coolant flows through these passages to carry off the heat. In the flathead V8, there are three feet of hot exhaust passages heating the coolant in the block.

When flatheads were built, water was the usual coolant, with additives such as alcohol for the prevention of freezing in winter weather. That's why the temperature gauge in many, but not all, shoebox Fords is designed to peg itself full hot at 212 degrees, the boiling point of pure water. With the advent of 7-pound radiator caps in 1951, new sending units were supplied to adjust the maximum gauge reading to about 230 degrees, a more accurate measure of the boiling point.

Today, we use a water and coolant mix. The reasons are simple: First, ethylene glycol-based coolant absorbs more heat than water does. Second, by using a 50/50 mix of coolant and water, we can raise the boiling point to 225 degrees or so. Add a good 7-pound radiator cap, and you're up into the 230-plus degrees boiling point range. Every pound of cap pressure raises the coolant boiling point by 3 degrees. In addition, using a 60 or 65 percent mix of coolant to water can increase the boiling point even more — but don't go beyond that percentage. Third, coolants contain anti-rust additives, corrosion inhibitors and water pump lubricants that water does not. Fourth, coolant is also an anti-freeze.

Final note on coolant: Most experts agree it should be changed every two years. The smart flathead owner will change it every year and use distilled water instead of tap water, because there are no heavy mineral deposits in distilled water.

And don't forget there are three important elements in an internal combustion engine cooling system: Coolant, radiator air flow, and engine oil. Ignore one of these and you'll develop a problem sooner or later.

THE FLATHEAD COOLING SYSTEM

The flathead cooling system is just like any other. It is composed of a radiator, water pumps, heater, belts, hoses, passages in the heads, the block temperature sending units, the dash gauge, and the radiator fan.

The radiator is the single largest component. If you're lucky, your car has one of the optional heavy-duty radiators (8BA-8005-B) Ford came up with in December 1949 to eliminate overheating. Ford Service Letters beginning in early 1949 are filled with tips for solving overheating problems, and news of optional radiators and fans. Thus, the flathead cooling problem has been with us for 40 years and no doubt will be for another 40 years.

But most cars were not equipped with the heavy-duty radiators. So the first thing to check is the radiator itself. Make certain it has the proper prefix on the upper right (passenger side) portion of the top radiator tank. Several years ago, Newsletter editor Mike McCarville was having trouble with his flathead in really hot weather. We worked on the cooling system for months before Mike accidentally discovered that the radiator was not marked 8BA, but carried a later 1952 Ford part prefix. We dug out an 8BA radiator, had it worked over at the radiator shop, and Mike's problem was solved. The '52 radiator has fewer cooling fins per inch. Yet all along, we had assumed the car had the right radiator. So check your radiator to make certain it is the right one for your car. Note: Maximum cooling is achieved through the use of radiator 1BA-8005-C and fan OBA-8602-B.

DOWNFLOW RADIATORS

Flathead radiators are of the downflow type. That means the water flows up and down in the coolant tubes. A drawback of this is that it allows sediment and other trash to settle in the bottom tank, which eventually will block the flow of coolant. Most modern radiators are of the crossflow style, which tends to eliminate this problem. If you're having cooling problems and haven't had a radiator shop go through your radiator, we suggest you do so as a first step.

WATER PUMPS

Flathead water pumps have been praised and cursed. Some say they are so good they pump too much coolant too fast to remove engine heat. Whatever the case, they do cause the coolant to circulate so they must be in good working order or you have a problem.

HEATER

The heater is also a part of the cooling system. It's nothing more than a small radiator that allows the use of warm engine coolant to heat the car interior. More than one flathead owner has turned on the heater in the summer to help cool an overwarm engine.

BELTS

V-belts, usually called fan belts, are obviously a vital part of the cooling system. If they are worn, cracked or slipping, they're not doing their job and should be replaced or tightened.

WATER HOSES

Water hoses, too, are an essential part of the cooling system. Regular replacement every two or three years (Gates Corporation recommends replacement every four years) can help ensure they never become a problem. Don't forget that hoses can look fine on the outside but be collapsing inwardly. And if your top hoses have springs in them, check those springs periodically to make sure they don't have a bunch of trash hanging on them. A friend of ours removed his one time and found a cigarette package wrapper tangled up in the spring!

WATER PASSAGES

The water passages in the flathead engine have a tendency, because of their design, to retain mineral deposits. This means that over time they will narrow themselves and could even become almost totally clogged shut. When this happens, it is time to find heads with clear coolant passages. If you have a persistent overheating problem and have eliminated every other possibility, suspect clogged passages in the heads.

TEMPERATURE SENDING UNITS

There's considerable confusion about the temperature sending units in the heads. The V8 has two. The single-prong unit actually monitors coolant temperature and transmits that electrically to the dash gauge. The double-prong unit is designed to peg the dash gauge at the boiling point of water (212 degrees) on 1949 and 1950 models. 1951 gauges peg at about 230 degrees. We have found that in some cases these head units are actually giving false readings, leading the driver to believe the engine is overheating when in fact it is not. As the head units age, the points become dirty, and electrical resistance increases, thus giving false (high) gauge readings. The head units should be matched to the thermostats and radiator cap to be completely accurate. A 4-pound cap 160-degree thermostat and a single-prong unit stamped 224 F should give an accurate reading at the dash gauge. With 7-pound caps, use sending units 1A-10884, stamped 233 F; and 1A-10990, stamped 226 F for complete accuracy. With this setup, the dash gauge pegs at about 230 degrees.

DASH TEMPERATURE GAUGE

The dash gauge used in the 1949, 1950, and 1951 Fords is electrical and pegs itself at either 212 degrees, or 230 degrees. A common question involves what actual temperatures the marks on the gauges represent. We used a mechanical gauge in one head and the stock gauge in the other to monitor engine temperature at each mark on the gauge in a '50 model. The results show that engine temperature, when the stock gauge is at the full cool mark, is 100 degrees or less. When it is at the first mark (which could be called the quarter mark), it is about 120-130 degrees. When it is at the second mark, or halfway point, it is about 156-165 degrees. When it reaches the third, or three-quarter mark, it is about 175-188 degrees. Halfway between the three-quarter and full hot mark, it is about 192-195 degrees. Pegged, it is about 210-226 degrees. With 7-pound cap and the proper head unit, the gauge indicates 230 degrees at full hot.

We've given approximate temperatures, because we know from experience that the gauge readings will vary from car to car.

RADIATOR FAN

The radiator fan has one role in life, to suck air through the radiator and cool the coolant while the car is sitting still or moving very slowly. At other times, forced air flow is usually enough to keep the coolant below the panic level.

The standard 1949 Ford fan had four blades. In August 1950, a 5-blade, 18" fan was made available as part number 8BAS-8602 for the '49 V8 only. In March 1951, the 5-blade, 18-1/4" 0BA-8602-B fan was available for '50-'51 models. The heavy-duty 1BA-8005-C radiator was suggested for '50-'51 models with overheating problems. FoMoCo soon decided the fan made too much noise, so it was replaced with the 3-blade fan used on most 1950 and 1951 models. If you suspect that a cooling problem is being caused by low air flow through the radiator at slow speeds or when stopped, you can bend the steel blades to increase their sucking power (which we don't recommend), and in the process you'll probably increase the noise the fan makes and disturb its dynamic balance. You'd be lucky to find one of the optional 5-blade fans or heavy-duty radiators, but another alternative is to install a fan from a '50-'53 Mercury, which featured a standard 4-blade unit. The Merc fan is wider and has more blades than the standard Ford model, which is a contributing factor to greater air flow through the radiator. And that means a cooler running engine.

TROUBLE SHOOTING

Despite the simplicity of the flathead cooling system, trouble shooting can be anything but simple. There are some common causes of overheating, and they are as follows:

1. Stuck thermostat(s)
2. Hoses collapse inwardly
3. Low coolant
4. Blocked radiator fins
5. Dirty or clogged radiator
6. Low engine oil level
7. Shot water pump(s)
8. Dragging brakes
9. Ignition timing too far advanced or retarded
10. Clogged heater core
11. Blocked or kinked exhaust or tail pipe
12. Loose cylinder head bolts

If you have a coolant loss problem, there are also some common causes you'll want to check.

Visible Leaks:

1. Loose hose clamps
2. Leaking hoses
3. Thermostat housing leak
4. Radiator core leak
5. Bad radiator cap
6. Leaking radiator core

Mysterious Leaks:

1. Blown head gasket
2. Cracked head or block

Most flathead overheating problems can be remedied by a step-by-step approach to the problem, systematically going through each component of the system and eliminating it as the cause.

A FEW COOL TIPS

There are some things you can do to help ensure you won't have an overheating problem.

1. If rebuilding an engine, carefully check the water ports around the valves and everywhere else. They can become clogged with rust or scale. Clean the ports with whatever you can find that will do the job — a clothes hanger, if all else fails. Check the passages in the heads as well. If they're clogged shut, find some new or better heads.

2. Paint the radiator with flat black paint (do not use gloss black). This helps remove heat. But don't lay on four coats of flat black. One light coat is plenty, since a paint build-up will be counter productive and insulate the heat instead of helping to remove it.

3. If you've been using multi-viscosity oil of 10w-30 or 10w-40, switch to a slightly heavier grade, say 20w-50. Likewise, if you've been using straight 20-weight, try 30-weight. Oil is an important part of the overall cooling system, since it reduces friction and removes friction heat as well.

4. Don't assume your coolant is good. It deteriorates just like anything else. Replace it regularly.

There you have it, everything you need to know about the flathead cooling system in order to keep it running efficiently and keeping your flathead running cool.

by Ken Mathias

SHOEBOX FORD/MERC

OVERDRIVE TRANSMISSION

If you're doing a shoebox Ford that has an overdrive transmission, there may be some apprehension about retaining the overdrive. Don't be afraid. The chances are good that the transmission you have can be used with little or no work, and if there are problems, they can usually be corrected using common tools and careful attention. The overdrive transmission is simply a common 3-speed gearbox with an attached tailshaft housing containing additional gearing and controls. The unit offers the advantage of giving the shoebox increased off the line performance, as well as providing the option of better fuel economy and less engine noise at cruising speed. With the overdrive locked out, engine power is fed directly to 4.11:1 rear axle ratio gearing. Engaging the overdrive reduces engine rpm 30% at the end of the transmission, thus giving an effective rear axle ratio of 2.88:1 (4.11x.70). Once engaged, a downshift to second gear can be made, giving a combination that is highly effective for mountain driving. So, there are advantages to keeping that overdrive.

Working on the 3-speed gearbox is the same as on any non-overdrive transmission. Work on the overdrive is just more and different. It is highly recommended that a service manual be procured and studied before tackling any work on the transmission. Reproductions of original manuals are available, and you may already have one for your project car.

For the purposes of this session, we will approach the subject in a step-by-step manner, from the most basic theory of operation, to troubleshooting, to a full-fledged overhaul. Information given is not detailed, but is intended to provide a general summary of procedures, along with a few tips that may not be in the service manuals.

As in any automotive servicing, do not go beyond your capabilities. Engines and transmissions are heavy, awkward, and can be very dangerous if

Transmission side cover removed, and 3-speed box pulled away from overdrive unit, showing relative positions of transmission gearing, overdrive gearing and overdrive housing.

Unit in upright position, with gearing in place, ready to receive overdrive housing.

Assembled and wired, complete transmission and overdrive assembly are ready to be reinstalled in the car.

mishandled or improperly supported or moved. Vehicles must be supported with jack stands of commercial quality. Replacement parts must be of a quality equal to or better than original. Do not attempt to replace fasteners with ordinary low-strength items. The use of stainless steel fasteners to improve looks in areas requiring critical strength must not be done unless you are qualified to calculate the stresses and specify the correct stainless fastener.

THEORY OF OPERATION

The overdrive unit consists of planetary gears, an overrunning clutch, and mechanical and electrical control. Mechanical and electrical events must happen for the overdrive to work. The overdrive is operated by first pushing forward on a control cable located under the dash. That action moves the shift rail lever, located on the left side of the overdrive unit, and changes the position of internal planetary gearing. At a speed of 27 mph, electrical contacts in a governor (located on the overdrive) allow a relay (located on the firewall) to operate a solenoid (located on the overdrive). The solenoid pushes a pawl toward the planetary gear arrangement, and the overdrive engages when the accelerator pedal is slightly released, and the pawl engages. A passing or kickdown feature is incorporated. Full depression of the accelerator pedal activates a switch located under the pedal, which turns off the solenoid with a simultaneous shut off of the ignition system (for about one half engine revolution), releasing torque on the solenoid pawl, allowing it to pull out and thus deactivating the overdrive. Reducing speed below 21 mph electrically and automatically disengages the overdrive.

TROUBLESHOOTING

The most effective way to determine satisfactory operation of the overdrive is in an operating vehicle. Given this ideal situation, the procedure is to start the engine, push in the operating cable, accelerate past 27 mph (at which speed the click of the relay should be heard), let up on the accelerator and the overdrive should engage.

If this doesn't happen, there are two possible locations of trouble: Electrical or mechanical. The most likely is electrical. With the cable pushed in, the car should freewheel, or coast freely, when letting up on the accelerator. If this happens, the problem should be electrical. Check the wiring connections for looseness, and correctness between the governor, solenoid, kickdown switch, relay, ignition coil, and main power source. A fuse at the firewall-located relay could be bad. If replacement of the fuse doesn't cure the problem, begin a methodical check of each component. The governor, kickdown switch, and relay usually stay healthy enough to operate the system. The standard problem lies in the juice can shaped solenoid located on the side of the overdrive. It is removed by taking out the two retaining bolts and turning 1/4 turn to disengage the plunger. Be warned that oil will come out of the plunger hole. The plunger is sealed with a readily available seal, which is a good practice to replace. Use a volt/ohmmeter to check the coil as well as the other electrical parts and circuits for open or short conditions.

If parts are needed, note that overdrive units were used on many brands of cars from 1949, and used parts can be obtained from salvage yards. My Ford has a $5 solenoid taken from a Studebaker. Make sure the electrical parts (relay and solenoid) you obtain are of the correct voltage for your application. The service manuals provide more detailed diagrams.

If the car does not freewheel, the problem is mechanical. Be sure the cable is connected to the overdrive shift rail such that full movement occurs. Beyond that, the problem lies internally and we're heading toward disassembly.

Evaluation is less precise, but still possible, if your transmission is lying on the floor. Manually turn the input (engine side) shaft while moving the transmission levers, and note the relative rotation of the output shaft. Reverse should be obvious, with Low and Second gear positions giving gradually more relative movement at the output end. High gear will give the same movement out as in. Moving the overdrive shift rail lever clockwise will mechanically engage the overdrive and enable you to check for freewheeling. It can be felt by manually turning the output shaft, if the overrunning clutch engages in one direction versus the other. Inability to turn the input shaft with your hands signals possible mechanical trouble.

Testing the electrical system out of the vehicle can

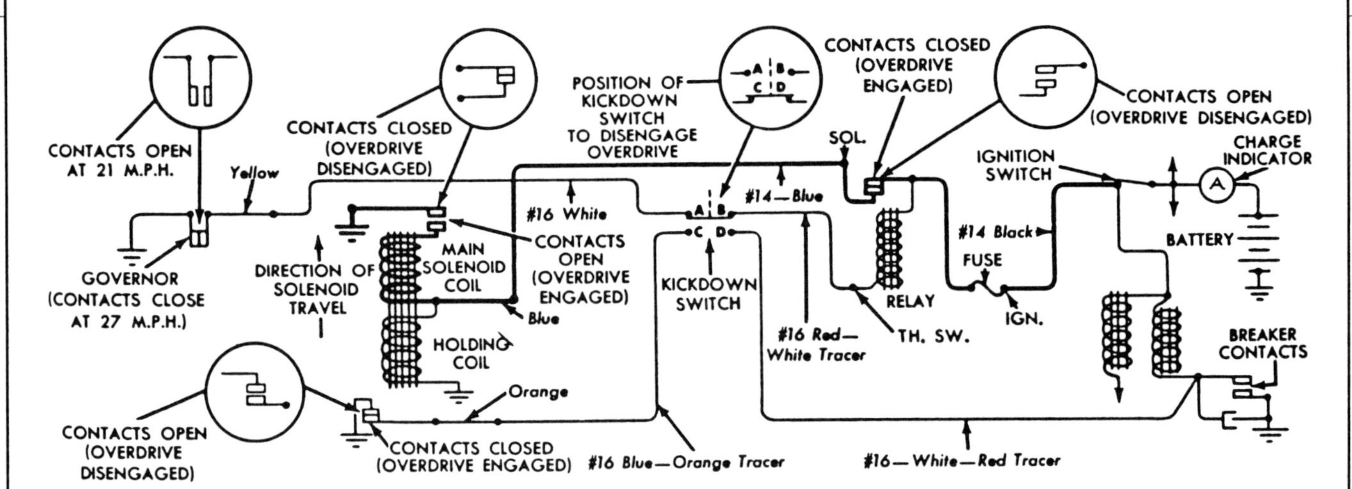

Overdrive electrical circuit indicates how the wiring is routed, and what takes place as the contacts open and close during operation of the overdrive unit.

be done, but it is most easily accomplished in an operating vehicle.

It goes without detailed explanation that if unusual noises are evident during operation, things are not mechanically correct.

IS OVERHAUL NECESSARY

The standard transmission/overdrive units are rugged and long lived, especially since most were used behind relatively weak engines, which got a lot weaker as they aged. My own overdrive had well over 100,000 miles and had been run low on lubricant at least twice when I was regularly driving it. Disassembly indicated no noticeable wear, and the overdrive parts looked factory fresh. Other than installing fresh gaskets and seals to cure the leaks, the only part replaced was the tailshaft bushing, and it was not worn enough to cause problems. If things seem to work okay, a good cleaning of the outside, and painting/detailing could put you on the road.

DISASSEMBLY

If you have to get into either the 3-speed box or overdrive, it's a good idea to do everything at the same time, mainly to install new gaskets and seals. After removal, do a thorough cleaning of the outside. By this time, the crud on the outside has nearly turned to stone, and wire brushes, power brushes, and scrapers are necessary. Once scraped and brushed, limited amounts of solvent can be used to finish it up. These days we are discovering that small amounts of solvent can pollute a lot of good ground water, and limited use and careful disposal of these materials can help the environment as well as our pocketbooks.

The best tip one can offer in transmission work is to pay attention. Start with the transmission lying on a clean work surface, ideally twice as long and wide as the transmission.

Remove the transmission case (side cover) bolts. Lift off, holding the shift forks in position during removal. Take note of their location and position. This procedure applies to all other parts removed. Place the removed parts in an orderly position on your work surface to help keep you and the parts correctly oriented. Drive out the pin which locks the Reverse and counter shafts. Drive the countershaft cluster gear shaft out from the front with a piece of 3/4" shafting just the same length as the countershaft. This shaft will stay in the cluster gear and retain the 44 needle bearings. Alternatively, a brass drift can be used, but the needle bearing rollers will be released from each end of the countershaft. With luck, they will stay inside the housing. Keep track of the thrust washers and note their original positions. This is where having a clean work area, preferably with a retaining lip, pays off.

Continue with the disassembly, carefully noting the position of everything. The transmission mainshaft passes into the overdrive unit and cannot be removed until later. Check the gears for excessive chipping, the bearings for roughness, and everything else for breakage and excessive wear. The main area of wear in a transmission is at the thrust locations. Make sure play is not excessive, and if so, obtain and fit additional thrust washers. Drive out the pins holding the shift cams in the side cover and replace the O-ring seals. The pins may be difficult to remove. If so, to avoid breaking the side cover, drill the pins out. New ones can be obtained or made from scrap stock (nails) turned in an electric drill.

Disassemble the overdrive unit. Start by removing the solenoid as described above. Remove the governor, which screws out, and is behind the solenoid. Although a special tool is called for, slip joint pliers (padded with several layers of cloth) applied at the base of the part should work. Remove the four bolts which retain the overdrive housing to the adapter plate. Do not remove the bolt which holds the adapter plate to the

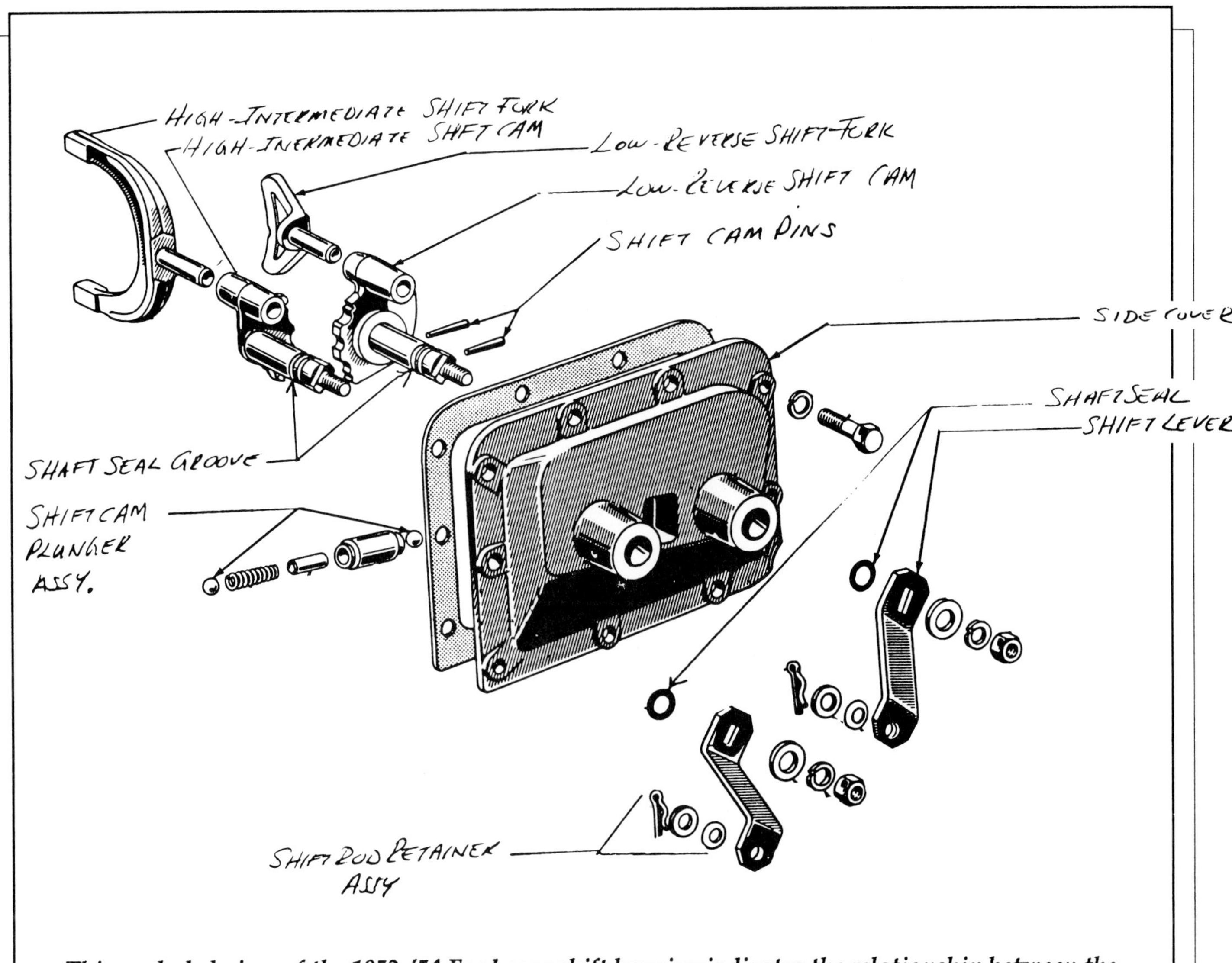

This exploded view of the 1952-'54 Ford gear shift housing indicates the relationship between the levers, forks, pins, shift cam plunger assembly and other components.

transmission. Drive the shift rail lever retaining pin out (the pin which retains the control shaft and lever on the left side of the unit). Slide the shaft out. Remove the small cover on top of the housing. Using snap ring pliers, screwdrivers, or long nosed pliers, spread the retaining ring, tap on the overdrive mainshaft end with a soft hammer, and slide the overdrive housing off, leaving the shaft in place. Pull the mainshaft off. Twelve rollers from the overrunning clutch will drop out, so be prepared.

You have now accessed the inside of the overdrive. Remove the U-shaped clutch retainers and slide off the clutch, planetary gear assembly, sun gear, and shift rail. Be aware the shift rail coil spring may fly out. Work slowly and carefully. Take out the large circular snap ring and remove the balk ring, plate and trough, sun gear and pawl. Mark (don't use a punch) or otherwise note the position of these parts — it will help prevent a lot of sweat at reassembly time.

You will now be able to see how the transmission mainshaft is removed. Examine the planetary gears for chipped teeth, and excessive wear. Be sure the inside of the overrunning clutch drum is smooth and not worn or "chattered." Inspect the balk ring for proper tension as outlined in the service manuals (it will turn one direction easier than the other). Replace the solenoid plunger seal and the shift rail lever seal. Remove the rear overdrive mainshaft seal and drive or cut out the bushing. Clean the inside surface, apply some anti-seize compound and carefully drive in a new bushing, followed by a new seal.

ASSEMBLY

After cleaning, painting the cases, and obtaining any new parts, it's time to reassemble. Use a commercially available gasket set and recommended sealants. Pass a large fine-toothed flat file across the flat surfaces which will receive gaskets, to remove paint overspray, burrs or other imperfections which

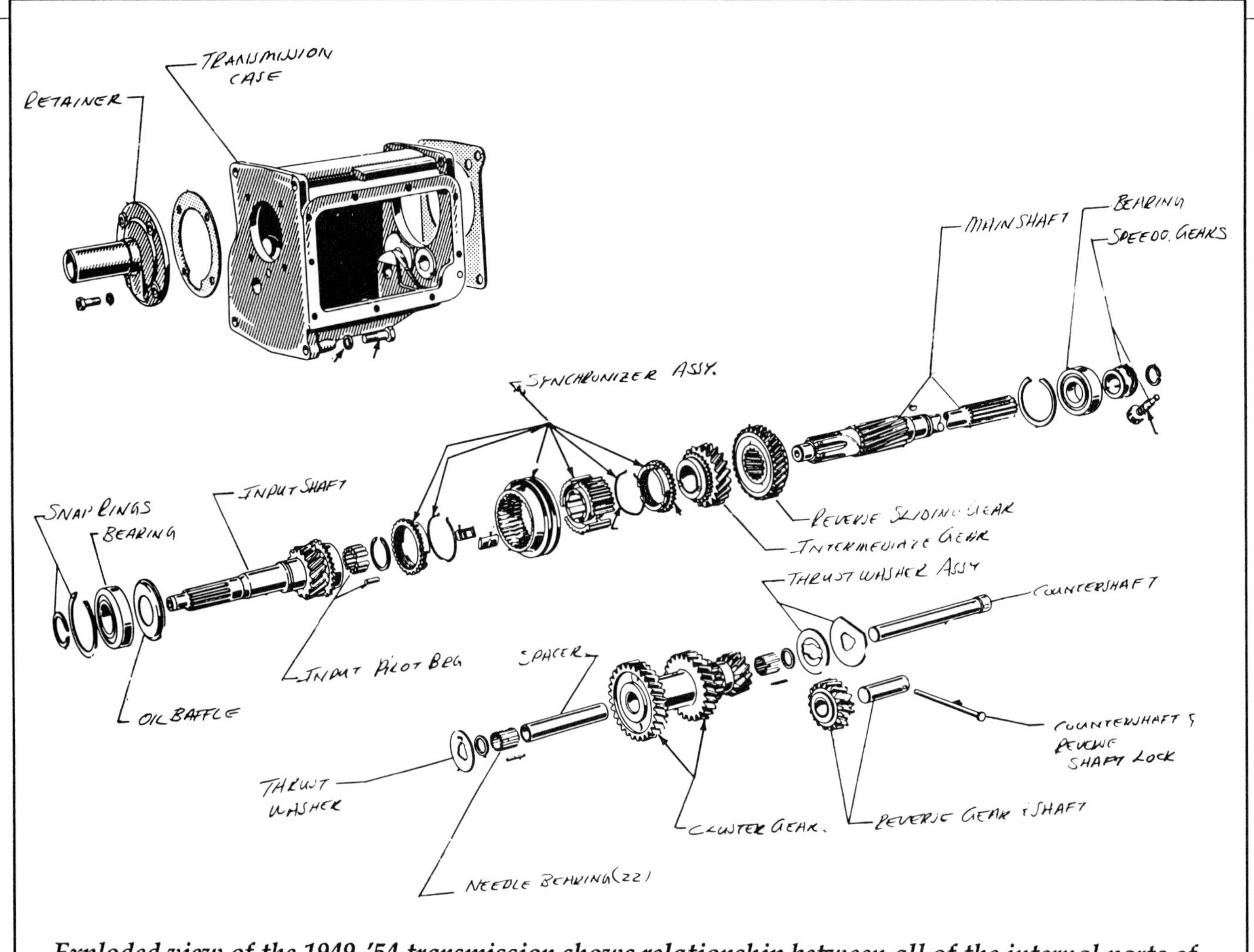

Exploded view of the 1949-'54 transmission shows relationship between all of the internal parts of the assembly.

could interfere with sealing. The technique is to lay the file on the surface and move it around with the fingertips, keeping the file flat at all times, and not concentrating at any one location. Check to be sure everything is clean. Lubrication with transmission oil during assembly is recommended. I like to use small squirts of aerosol motorcycle chain lube. It's more convenient than oil, less messy and smelly, and used in such small amounts it should not cause any adverse reaction with the gear lube.

Assemble the 3-speed box in reverse order of disassembly. Use wheel bearing grease or other stiff grease to "glue" the rollers to the inside of the countershaft gear cluster. Insert the piece of 3/4" shaft cut the length of the countershaft gear cluster. This will hold the rollers in place. When the countershaft cluster and thrust washers are in place in the box, drive the countershaft in from the back, displacing the temporary shaft. Make sure the locking pin hole in the countershaft is in the correct horizontal position while driving it in. When all of the 3-speed section is in place, work the parts around to be sure everything slides and moves freely. Tap the countershaft and reverse shaft locking pin in, applying a dab of sealant to the ends. The side cover and shift forks should be temporarily installed to check for proper action.

Assemble the overdrive section, also in reverse order. Work the parts as you assemble them, to be sure they are in the proper position with no binding. After the clutch assembly is snapped in place, turn the transmission upright. Use a stiff grease to "glue" the 12 overrunning clutch rollers in place. Install the overdrive mainshaft over the clutch and rollers. Then, slide the overdrive housing downward over the mainshaft. Spread the retaining ring in the housing until the housing goes into final position. Getting the ring spread just right sometimes takes awhile, so be patient. Attach the 4 housing bolts. Push the shift rail lever in from the side and install the retaining pin. Work the lever and it should move freely, with the action of the spring being felt. Finally, install the solenoid and governor. Install the transmission side cover assembly, using sealant on the bolts. Turn the

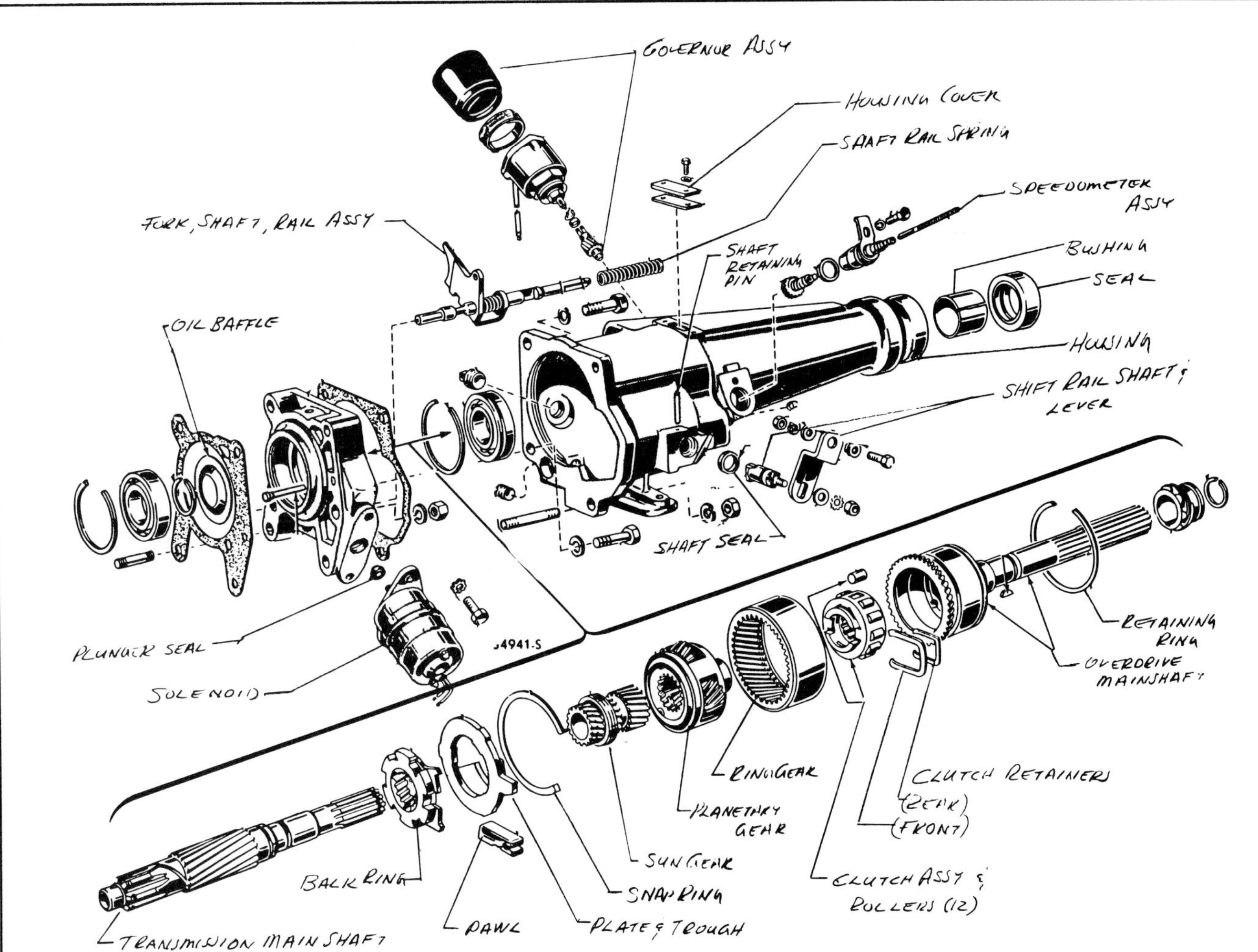

This exploded view of the transmission overdrive for 1952-'54 Fords gives a clear picture of how all of the internal components are related. Very helpful information during reassembly.

input shaft and run through the gears, as in the Troubleshooting section above. When you are sure everything is correct, tighten all fasteners to the proper torque. Sealing of the shafts which are driven through the 3-speed case is reliant on the tightness of fit. A way to help eliminate leaks at those points is to apply a smear of either hardening or non-hardening type sealant over the ends of the shafts after installation.

HOOKING IT ALL UP

Previously, you may have had your flywheel resurfaced, new ring gear and pilot bushing and a new pressure plate installed. A top quality job will have these parts, as well as engine parts, balanced. Use a new clutch release bearing. Make sure the pressure plate release levers are adjusted to be exactly even when the pressure plate and clutch disc are bolted up. Experience is that they do not come adjusted correctly when new or rebuilt. Attach a dial indicator or a piece of reference wire to the engine block and adjust the 3 lever screws while turning the flywheel. If a different engine/housing/transmission combination is used other than ones you know were factory-mated, check that the clutch and flywheel housing is aligned. Shimming may be necessary. Service manuals outline these procedures. A bit of additional checking of the work at this point will help assure a smooth shifting vehicle, and a longer lasting transmission and clutch.

Using guide pins on the 2 lower bellhousing transmission attachment holes (3" bolts with the heads cut off), slide the transmission in place. Install bolts and torque to the factory specification. After installation in the vehicle, hook up the linkages and wiring. It may be necessary to build up or bush some of the linkage connections to compensate for wear. A common point of wear in the Fords is at the steering column linkage. Wear on the column tube pin and levers can be corrected by fitting an oversized pin.

When the driveshaft is in place, fill with gear lubricant through the main transmission filler plug. The overdrive lubrication is common to the 3-speed gearbox reservoir.

by Skip Readio

6-VOLT TO 12-VOLT CONVERSION

For various reasons, it often becomes necessary to convert 6-volt automotive electrical systems to 12-volt operation. Years ago, it was to allow for the use of the higher voltage for improved starting. Today, it has become necessary due to more modern engines being transplanted into older cars.

Converting the charging system, or changing the whole power plant, requires that the rest of the vehicle's electrical system be looked at.

• All lamps must be upgraded to 12-volt units.

• All instruments have to be isolated (power routed through a voltage reducing device) or replaced.

• Ammeters will have to have polarity reversed, because nearly all 6-volt systems employ positive ground, while 12-volt systems use a negative ground. So, the ammeter will read backwards if it's not rewired. (Don't worry, this is a VERY simple task to accomplish in most cars.)

• The heater motor must be isolated by a voltage reducing device, or replaced with a 12-volt unit.

If you're converting to 12 volts to simply improve starting characteristics, you'll have to change the ignition coil and install a ballast resistor in series with the coil and ignition switch (between the ignition switch and the coil).

Another consideration is that the factory radio won't work any more, because it's a positive ground device and you've just converted to negative ground.

There is one more alternative to improved starting. Many owners have used 8-volt batteries to improve starting, because none of the costly or bothersome component upgrades are involved. However, there are a few problems with this approach. First, you have to adjust the voltage regulator to put out 8 volts. This isn't an easy job. Yes, you can bend the tab and get the generator to charge higher, but to do it correctly requires (1) an ammeter be placed in the circuit right at the generator, (2) the battery to be fully charged, (3) a voltmeter regulator be placed across the battery terminals, and (4) the voltage regulator be up to operating temperature. Most of these voltage regulator adjustment procedures are covered in older "Motors" manuals, and the like.

After spending all the time to do this, the generator isn't capable of putting out enough voltage to charge an 8-volt battery. It requires nearly 9 volts to charge an 8-volt battery. While spending all this time trying to adjust the regulator, you're gradually wearing down the battery. How do you charge it back up again? Well, you can try a 12-volt charger, but I wouldn't recommend it unless something like a headlight bulb is wired in series with the positive charger cable to drop the voltage a bit. Put a voltmeter across the battery and make sure the voltage stays around 9 volts. Keep watching the voltage, because it'll change as the battery charges. Maybe you're starting to get the picture here. Charging an 8-volt battery is a real nuisance.

Another real nuisance is finding an 8-volt battery, especially when you're on the road. Farm Bureau stores are a good bet, as are honest-to-goodness battery shops. Auto parts stores will most likely have to order one for you.

If you only use the car locally or for reasonably short periods of time, go ahead and install an 8-volt battery. Adjust your voltage regulator to keep the battery charged as best you can. Put together a special hookup for the battery charger to top off the battery charge when you get back home. You'll probably have to experiment a bit with different combinations of resistors, depending on the brand of battery charger you have, to get 9 volts out at the end of the wires, but it should be fine once you've worked out the variables.

If you plan on driving the car for any length of time between charging times, switch to 12-volt electrics. The first thing we'll cover here is what you have to do to the rest of the car to get it to work with a 12-volt system. Then, we'll get into what you have to do under the hood to either convert just the charging and ignition system, or to convert to a later model motor with its associated electrical system.

To start off, all lights will have to be replaced.

Headlights, parking lights, license plate lights, dome light, dash lights, trunk light, under-hood light, courtesy lights, glove box light, etc.

Then, reverse the polarity of the ammeter. This is accomplished by removing the wires from the battery side of the ammeter and placing them on the generator/alternator side of the ammeter, and taking the wire from the generator/alternator side and putting it on the battery side. Only one wire should be on the battery side of the ammeter. This 10-gauge wire will connect to the starter solenoid on the same 3/8" stud as the positive battery cable. All of the other wires (10-gauge from the alternator or generator voltage regulator, 12- and 14-gauge wires to light and ignition switches, etc.) should be on the other ammeter terminal.

Most vehicles are constructed with ammeters having 10x32 studs on the rear. Some, however, employ an ammeter wherein the 10-gauge wire from the charging device to the battery just passes through a loop on the back of the ammeter. These are a bit more difficult to convert, as one half of the wire has to be removed from the harness so that it can be pulled back through the ammeter loop and fed back again in the opposite direction. This is the preferred way of reversing the polarity.

The wire can be cut and spliced, however, a soldered splice is recommended over a yellow crimp connector. Yellow crimp connectors are used for 10- and 12-gauge wire. Unless you have an expensive set of industrial crimping dies (they cost upwards of $130), I wouldn't recommend using a crimp connector in the charging circuit. Solder and shrink tubing or tape will make a much better connection.

VOLTAGE REDUCERS

Six-volt heater motors can be used, if a ceramic encased voltage reducer is placed in the wire feeding the switch or motor. If the heater switch has a lamp inside, placing the voltage dropping resistor between the ignition switch and the heater switch will allow you to retain the 6-volt bulb in the switch. Placing the resistor between the heater switch and the heater motor necessitates using a 12-volt bulb. It is worth pointing out here that 6-volt bulbs are getting a bit difficult to find in many corner auto parts stores, so the latter alternative might be the better of the two.

Six-volt windshield wiper motors and 6-volt electric fuel pumps can also be operated via one of these ceramic encased resistors.

INSTRUMENT CLUSTER REGULATORS

Instruments should not be isolated with a ceramic encased voltage reducer. Because the output of these voltage reducers isn't sufficiently stable, a more precise means of voltage regulation should be employed.

Ford Motor Company and Chrysler Corporation have used instrument cluster regulators for years, to reduce voltage in their 12-volt cars to approximately 6 volts for their instruments. These ICRs (sometimes called IVR - instrument voltage regulator) are found on the back of the instrument cluster in most pre-computer-dash FoMoCo and MoPar vehicles.

ICRs/IVRs are used to regulate fuel gauges, oil pressure gauges and coolant temperature gauges. They are not used on the ammeter or voltmeter. You won't find one if there is a low fuel indicator wired into the fuel gauge, however, because the manufacturers rely on the low fuel bulb for the voltage regulation.

The ICR/IVR is a little (approximately 1" long by 1/2" wide) metal can with three contacts that plug into the back of the instrument cluster. One pin is part of the metal enclosure (or is connected internally to the enclosure) and connects to chassis ground via the instrument panel and/or harness. The other two are battery voltage in, and regulated voltage out.

Look for a vehicle with the same number of instruments as your vehicle, because ICRs/IVRs are designed to regulate a specific number of instruments, and one designed to operate a single instrument will provide the wrong voltage to three instruments. When you figure out which late model vehicle has a similar instrument cluster, head on down to your local NAPA or other parts store and they'll have a listing for the ICR/IVR for that vehicle. Most ICRs/IVRs are imprinted with "in" and "out" or similar indications on the two active pins. The third pin, as mentioned, must be grounded.

Idiot lights should be converted to 12 volts, and late model 12-volt senders should be installed in the engine block.

ELECTRONIC VOLTAGE REDUCERS

Another means of reducing the voltage to the instrument cluster is to install an electronic voltage reducer. These devices provide a more precisely controlled voltage to the instruments and aren't load-dependent. In other words, the number of instruments controlled by the regulator has no effect on the voltage output to those instruments. Both Ron Francis Wire Works and Radio Shack market a pair of voltage reducers that will adequately regulate the voltage to your instruments.

The Ron Francis voltage reducers are designed specifically for installation in automobiles to control 6-volt equipment. One is rated for instruments only, while the other is powerful enough (15 amps) to control an electric fuel pump or something like a '55 Ford factory air conditioning solenoid.

Radio Shack units are designed to provide a range of voltages between 3 and 9 volts. The low-power unit (270-1560) is designed to plug into a cigar lighter and, while it is capable of 3-volt, 4.5-volt, 6-volt, 7.9-volt, and 9-volt operation, it can be wired into the system without using the cigar lighter, and set to 6 volts for powering instruments. The higher-powered Radio Shack unit (270-1562) provides only 6-volt or 9-volt outputs and is designed to be wired directly into the electrical system. While it doesn't have a current rating as high as the larger Ron Francis Wire Works model, it is quite adequate to handle vehicle instruments. I would recommend the latter model simply because it is a more compact unit and is less confusing from an installation standpoint, because you don't have to disassemble the cigar lighter plug and ascertain which wire is connected to the battery (the one connected to the center pin, usually fused as well) and which is connected to ground (the one to the side strap).

To connect any of these voltage reducers into your instrument circuit, first disconnect the wire that runs from the ignition switch to the instruments. The easiest place to disconnect the wire is usually at the ignition switch itself, because the wire is normally doubled up in the connector where it connects to the first instrument. The second wire in the connector runs to the next instrument and so forth. Between the ignition switch and the disconnected instrument wire is where the voltage reducer is installed. Follow the instructions packaged with the voltage reducer to get the proper colored wires on the proper points. Essentially, you'll have power and ground on the "in" side as well as power (to the instruments) and ground (to the dashboard) on the "out" side.

If original instruments are to be used with a later model motor, it will be necessary to adapt the original (or correct new replacement parts) coolant temperature and oil pressure senders to the new motor to ensure accurate instrument readings.

UNDERHOOD WIRING

The following paragraphs assume that you intend to utilize some or all of the existing vehicle wiring. If the intent is to totally rewire, conversion isn't the issue any more, and you should refer to Tex Smith's book "How To Do Electrical Systems" for guidelines.

To convert the engine from 6-volt to 12-volt operation, replace the battery, generator and regulator with a 12-volt version. Nothing else is necessary, as both generator systems share a common wiring principle.

If converting to an alternator, it will be necessary to slightly modify the charging system harness. You need not destroy any of the original wiring to convert to alternator charging. First of all, remove the battery, generator and regulator. Lengthen the wire that previously connected to the voltage regulator "BAT"tery terminal. This wire will now connect to the output of the alternator. If mounting the alternator in the same place as the generator was mounted, bolt the aforementioned "BAT"tery wire to the "ARM"ature, where these two wires formerly connected to the regulator, and connect the other end (the one that used to connect to the armature terminal on the generator) to the output of the alternator. This will leave one wire in the old regulator-to-generator harness. This wire can be used to connect the new alternator regulator to the field of the alternator.

For most alternator applications, it'll be necessary to run an additional wire up to the alternator as well as running another wire from the ignition switch to the voltage regulator to energize the regulator when the motor is running. Unlike a generator, an alternator must have an ON/OFF switch function associated with its operation. Even the single wire alternators have this function, however, it is automatic, thus negating the need for the extra wires to excite the alternator. For all conversions, however, you'll have to make the modification to the length of the "BAT"tery wire.

If you intend to install a later model engine, you can still modify the regulator wiring harness in the same manner.

When upgrading the operating voltage, replace the ignition coil and add a ballast resistor into the supply circuit for the coil. Remember, when hooking up the coil, the distributor will be connected to the negative side and the ballast resistor will be connected to the positive side.

On Ford products, it is necessary to change the starter solenoid to a 12-volt unit. It is a good idea to try to use a 12-volt starter as well. Many people have used 6-volt starters in 12-volt systems, especially in Ford flatheads, but long-term reliability suffers when the starter motor is driven with too much voltage for a few years. If your motor fires right up after one or two revolutions, the starter will last for quite a while. But if the motor is getting tired, and the starter has to grind for a few seconds, it's not going to last very long if you keep whacking it with 12 volts.

by Rich Johnson

FLOOR PAN REPLACEMENT

One of the harsh realities of working with cars that are four decades old is that time and corrosion have had a chance to do their evil work. Unfortunately, rust doesn't always limit its appetite to simple things like fenders and rocker panels. In some cases, metal cancer starts lunching on the floorpan, which is serious because the floor is an important structural member that gives strength and rigidity to the rest of the body.

In some parts of the country, it's difficult to find an original shoebox Ford that doesn't need a floorpan replacement. Once upon a time, that presented a major problem for car builders, but now there are aftermarket replacement pans available that make the job fairly easy. When we say fairly easy, we don't mean easy like replacing a piece of trim, but at least the job is simplified because you don't have to cut and trim and shape and weld a bunch of small pieces of sheetmetal to build a new floorpan. These pans are ready-made, correctly sized and shaped to fit.

Bradley Antique Automotive (4200 South I-85, Charlotte, North Carolina 28214; phone 704-392-3206) makes replacement pans for '49-'51 Fords, reproduced to stock detail. Bradley's pans include longitudinal deep grooves that Ford engineers originally specified be stamped in to assure greater rigidity and strength. Bradley searched far and wide to find stock pans that were suitable to use as patterns for making dies. He told us that this was one of the hardest parts of the whole project. There were plenty of good cars, but most had rotten pans. After locating a set of decent original pans and making the dies, it was necessary to build a special press capable of stamping the grooves in 18-gauge (.047" thick) cold rolled steel. From the accompanying photos, it's easy to see that the product is nice.

There are some installation tips that make the job easier. The first step is to brace the body internally to ensure that everything remains in correct alignment during the project. This is extremely important regardless of whether the body is totally removed from the frame or the pan is removed in sections while the body is still on the frame. If the body is not braced internally, when the floorpan is removed,

Starting with a fair example of a stock shoebox Ford floorpan, Bradley Antique Automotive created dies to be used in a special press to stamp out brand new replacement pans. One difference between the original and replacement pans is that the Bradley units don't include the driveshaft tunnel, which must be salvaged from the original floor and welded to the replacement pans. Shown is a stock '49-'50 pan. The '51 unit is somewhat different.

things can shift, and doors won't fit the way they were meant to.

Bradley recommends using scrap angle iron for bracing, tack welded in place for easy removal after the new floorpan has been installed. Measure carefully, and brace straight across, upright and diagonally. The doors may also be tack welded shut to serve as additional alignment support. Use plenty of bracing, because when the floor is removed the lower part of the body becomes quite flexible, and as little as 1/4" out of line can cause a bunch of trouble later on.

Factory spot welds need to be broken loose or removed to allow the floor to separate from the body, and several methods can be employed to accomplish this.

1. Grinding: Locate the spot welds and grind through the one layer of metal to be discarded.

2. Drilling: Using a 3/16" to 1/4" bit, drill each spot weld through only the piece of metal to be discarded.

3. Cutting: There are some places where it is difficult to grind or drill and where there is no joint. Best results will be achieved by using a hacksaw or sabre saw with metal cutting blade.

4. Chiseling: With either a hand or air chisel, drive between the two layers of metal to be separated, cutting the spot welds loose. Use this method carefully to prevent distortion and damage to the metal. After the floor has been removed, a chisel may be used to remove the small pieces left in tight places.

NOTE: It is not recommend to use a cutting torch, because of the resultant metal distortion and hazard of fire.

After removing the original floorpan, straighten the remaining flanges with a hammer and dolly, then grind smooth. The original driveshaft tunnel must be removed and reused, because Bradley doesn't offer replacement driveshaft tunnels, but this part of the original floor is usually the last and least to be damaged by rust. Once removed from the original floorpan, and cleaned of burrs and straightened if necessary, attach the driveshaft tunnel to the new floor pans with sheetmetal screws or small tack welds to allow for any adjustment before final welding.

If the project is a body-off, before welding permanently, it's a good idea to conduct a test fit with the body resting on the frame and all units tacked in place. At this time, alignment can be adjusted and holes can be drilled for body-to-frame bolts. When everything is perfect, final welding can be done along the joint between the driveshaft tunnel and the floorpan. Then, remove the body from the frame and finish welding the underneath side.

Avoid shortcuts. Take your time, follow the instructions, and the floorpan replacement project can produce excellent results.

This is an example of the dies used for making the replacement pans. This particular one is used to form the transmission area on the '49-'50 floorpan.

Blank 18-gauge sheetmetal stock has been nibbled in the transmission area in preparation for a trip to the press.

The finished product looks like this. Compare stamping with the stock floorpan, to get an idea of the kind of detail that has gone into the production of these replacement units. Shown are the full left and right '49-'50 Ford pans.

by Jim Clark

CARSON TOPS

This array of shoebox Fords and Mercurys began life as 2- and 4-door coupes or sedans. The Carson top treatment has converted them all into very desirable open air cruisers.

Many outstanding coupes and sedans have been built over the years, but the open cars have always been the first choice of most car enthusiasts. Cruising around in the warm sunshine in a roadster or convertible with the top down is an experience that is hard to top. This popularity has made them the most expensive models to buy or build though.

Converting a desirable coupe or sedan to a fresh-air machine is one alternative that has gained widespread acceptance in recent years. This is most easily accomplished by adding a Carson top, instead of creating a folding convertible top. The latter being an expensive and involved process.

Carson tops were created initially as a hardtop alternative for use on convertibles, but now are seen predominantly on open car conversions. They have come a long way in how they are constructed as well. The early versions had frameworks of tubing and wood covered with canvas and cotton padding. These were heavy and not very weather resistant.

Today's Carson top consists of a light fiberglass shell covered with 1/4" waterproof vinyl top padding and standard convertible top fabrics. Convertible top molded rubber seals out the elements, while over-center style marine hardware cinches them down in place. Opening or fixed rear windows are added, using aftermarket convertible units.

The relatively light weight of these modern Carsons lets two people install and remove them quite easily. Storage is not much of a problem either.

Slicing the metal roof off of a car can be an unnerving

At the rear corner of the roof, a cut is made along a line from the bottom of the rear window to the top of the door glass in the rear door. Plasma cutter in use here works best on single thickness material, as multiple panels cause damage to tip.

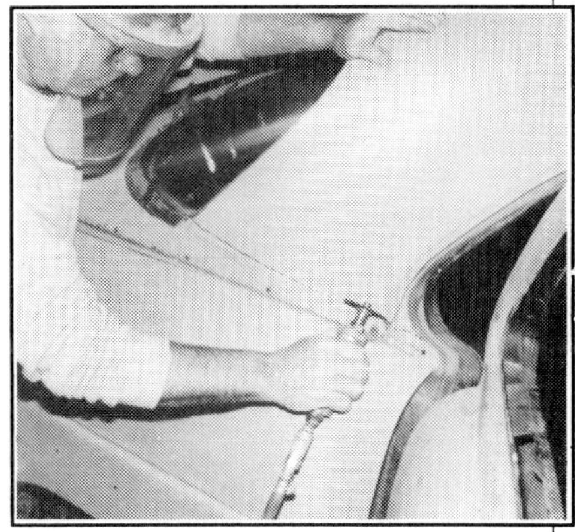

Cuts at corners where multiple panels exist are better done with a die-grinder or saw.

Door frames up front are cut slightly below window line to remove radius. Sheetmetal is added to straighten window line.

Rear door frames are cut in same manner as fronts. Circular saw with a carbide blade works well for these big multi-layer cuts.

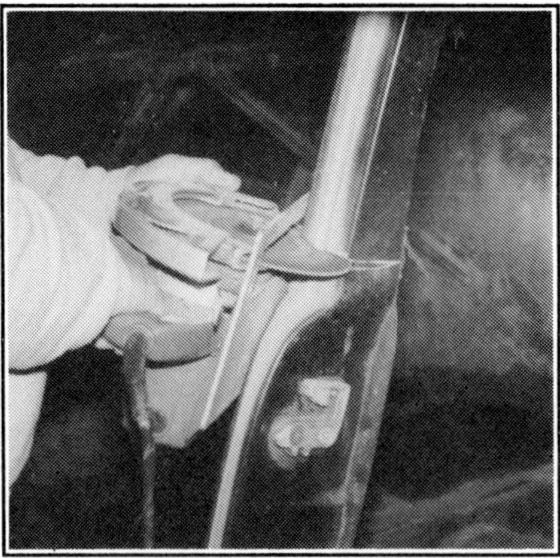

Circular saw is capable of cutting through B-pillar between doors. It's necessary to work in from both sides though.

Cutting the A-pillar at the windshield requires some planning before making the cut. A pattern of the new windshield should be made so curvature at ends can be controlled.

Top is cut free at A-pillar, using circular saw. Upper portion is cut from top to form upper edge of windshield frame.

Top windshield frame and sections of A-pillar are combined to create new opening. Welds show where cuts were made and trimming was done.

New cap from sheetmetal is fabricated and welded along top and rear edges of new windshield structure.

experience though, for one who has not done this type of thing before. To get the lowdown of how this should best be done, we enlisted the aid of Dick Dean, the master of top choppers. Over the years, he has dropped or removed the lid on more cars than anyone and developed the Carson top shown, for use on a wide range of makes and models.

In the accompanying photos, we have used one of the more popular models to illustrate the procedures followed in making the switch to an open air machine. The '49 thru '51 Mercs, both 2- and 4-door models, lend themselves to the treatment quite well.

Removing the roof doesn't weaken the car structurally to any appreciable amount, and finishing off the mating edges between top and body blends quite nicely. Chopping the windshield requires some careful cutting and fitting of the parts and frame, but the flat glass makes windshield replacement a very simple matter.

Side windows are also flat glass, so frames and vent window assemblies from Ford or Mercury hardtops or convertibles from the era can be modified to fill the openings. Chrome trim pieces for the door caps can also be adapted from these same models.

Another alternative to creating side windows is a much less expensive method; creating upholstered caps that fit over the tops of the doors.

When using the less expensive and more readily available 4-door as a subject, a common practice gaining favor has been to weld the rear doors shut, creating a long rear quarter panel. In this configuration, slide-in windows for the rear could be fashioned in place of the roll up type.

A review of the accompanying photos should give you an insight into the techniques applied to the task by Dean. With this new alternative, that old coupe or sedan that you saw and dismissed as not very desirable, may appear quite different to you now. Maybe they deserve a second look. But don't show this article to the seller, or the price may undergo some undesirable changes.

Vent window assemblies from '52 thru '54 Ford or Mercury hardtops are the most plentiful units available. Cut is made at top, creating a longer horizontal area by reforming frame.

Retaining the original windshield contours allows the use of stock trim in shortened configuration. Note small cuts in trim at center of A-pillars.

The amount of chop will determine how much comes off top of vent window. Bright trim piece from early '50s Ford hardtop provides finished look.

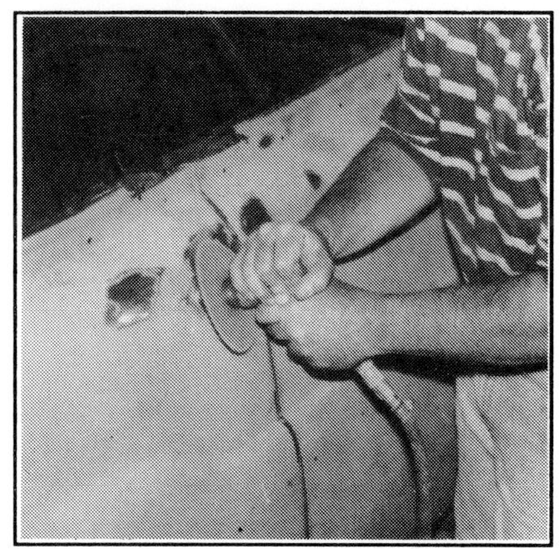

Area around door handles needs to be sanded to bare metal in preparation for filling.

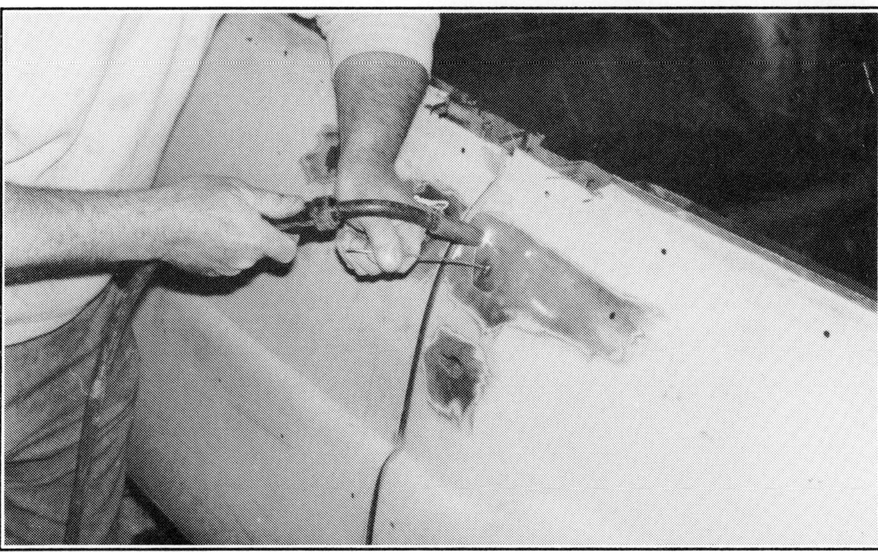

A neat trick Dean uses when patching holes is to tack rod onto patch, insert it in hole, and pull outward while welding. This eliminates the need for access to back of panel.

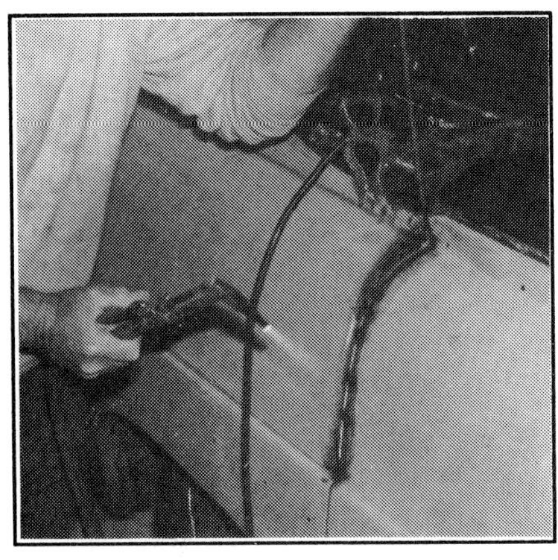

When rear doors are to be welded shut, a good way to avoid future cracking is to heat rod and form it into gap, tacking as you go.

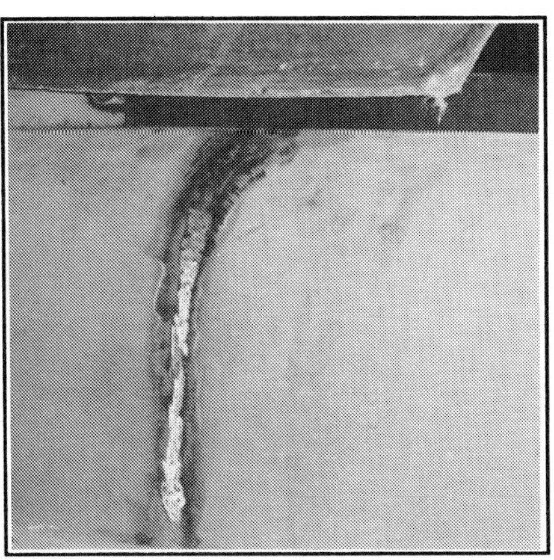

Welded seam is strong and will not crack out when car flexes. Long skirts can be used on new extended quarters.

Alternative to side windows is this U-shaped channel, fit over top of doors and secured with screws. Channel is upholstered before installation.

This 4-door has been given windowless treatment. Padded channel caps off top of doors, creating a true roadster. Vinyl trim continues around rear ledge area as well.

New top is one-piece fiberglass unit. Light in weight, yet strong and impervious to the elements.

Over-center latches pull top down against convertible top rubber seal attached at mating edges. Latches are available at marine supply.

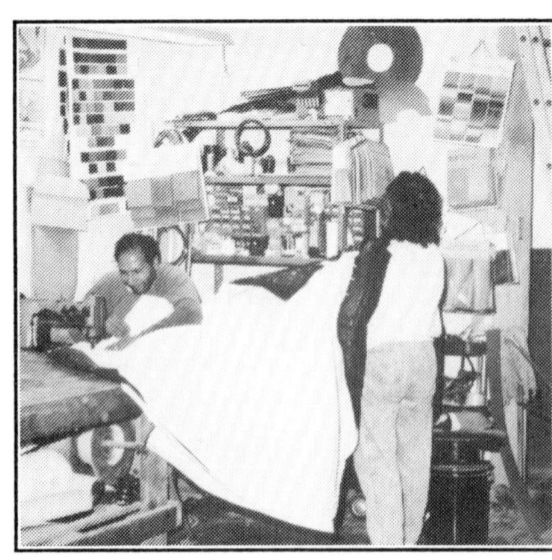

Covering for top is stitched, using convertible top material over 1/4" waterproof vinyl-top padding.

Material is stretched over fiberglass shell and glued in place. Headliner is applied to inside in same manner.

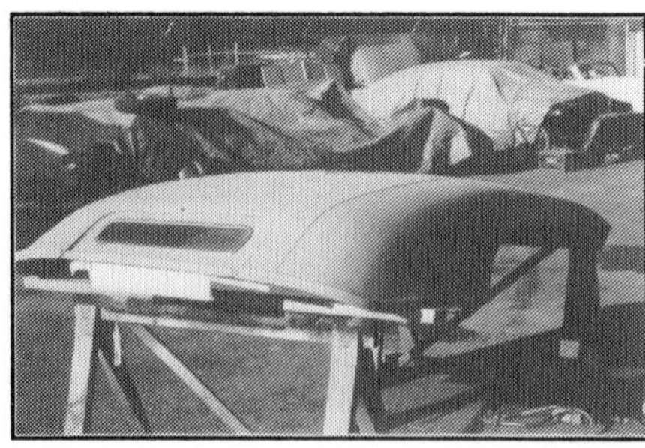

Finished Carson top looks like tight fitting convertible top. Rolled edge around perimeter of top helps divert moisture.

by Jim Clark

SHOEBOX FORD/MERC

A DASH OF MODERNIZATION

This machined aluminum panel and molded dash insert combine to update the old 6-volt equipped '50-'51 Mercury dash. Much of the original styling is maintained without the antiquated instrumentation.

Unless you are restoring a classic or special interest car, retaining some of the outdated mechanical or electrical components is inefficient at best. Bigger engines and modern drivetrains are usually the first areas to receive consideration. With many builders, this is where the modernization ends, overlooking the vehicle's nervous system, instrumentation and controls.

Updating these systems usually requires replacing everything with new components. The trend is to high tech, but sometimes these space age components can look very out of place in a traditional custom. The ideal compromise is a subtle blending of the two.

A blend of the high tech and traditional is achieved in this popular modification created by Dick Dean for the shoebox Mercurys. It features a fiberglass reproduction of the die-cast dash panel insert from the '50 and '51 Mercs, and a machined aluminum panel to house the gauges and controls.

The fiberglass insert fits into the stock steel dashboard and can be finished in a variety of ways. Among these are painting, either in complementary or contrasting colors, and covering it in leather or vinyl. Some trimming of the opening is required, to install the insert. Extra material has been left around the outside edges, allowing for slight variations in dash openings. Sanding with a disc sander is all that's needed to fine-tune the fit. Stock hardware holds it in place.

The instrument panel is precision machined on automated mills from 6061 T-6 aluminum. It can be polished, plated, anodized or painted to fit any desired customizing scheme.

Instrument sizes accommodated are 3-3/8" for the speedometer and tachometer, and 2-1/8" for the five smaller openings. These usually include water temperature, oil pressure, fuel, voltmeter, and clock. These sizes are for standard VDO gauges and are common in most other aftermarket lines.

Recesses machined into the rear of the panel position the gauges and facilitate mounting. Panhead screws engage the rear of the outer ring on the gauge, when installed into the pre-drilled holes, holding the gauges into the panel. This method provides a pretty unobstructed area behind the panel.

An opening, or individual holes, must be cut in the fiberglass insert for clearance behind the aluminum panel. Openings at the end of the aluminum panel are for air conditioner or ventilation ducts, but could be used for additional instrumentation instead.

Above the speedometer/tach openings are three small holes that house indicator lights for the turn signals and high beam headlights. Two slots at the right end of the panel are for the stock Mercury heater/defroster levers. A mounting hole for the light switch is provided at the left end of the panel.

Facilities for mounting a radio and speaker are lost in this conversion. Most of the new stereo systems demand different mounting facilities and have multiple speakers requiring that they be relocated anyway. Consoles from other late model cars, or custom made units, are usually the answer to solving this relocation problem. Consoles also provide a place for power window and power seat controls, equalizers, amplifiers, speakers and in some areas, mobile phones.

By switching the later sheetmetal dash panel, this

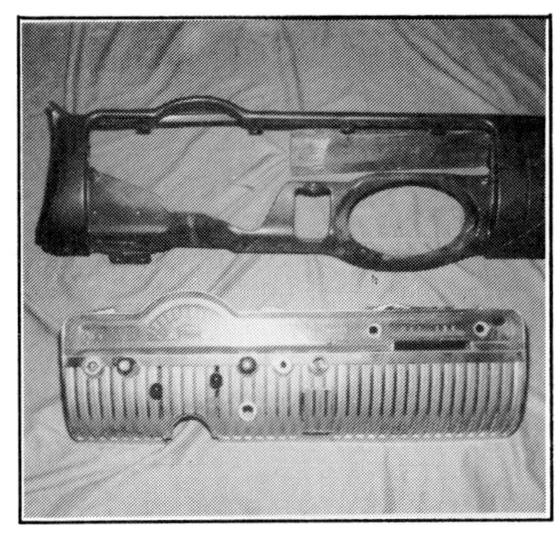

Stock dash is a two piece unit. A stamped steel panel that bolts into the firewall and a die-cast plated insert that houses the instruments and various other controls.

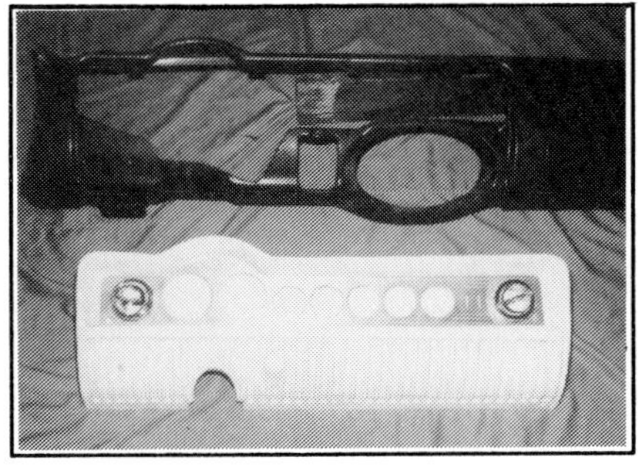

Replacement for the stock insert is a fiberglass replica which houses a machined aluminum panel pre-cut to accept a full set of instruments. Ducts at end of panel are heat or air conditioner outlets.

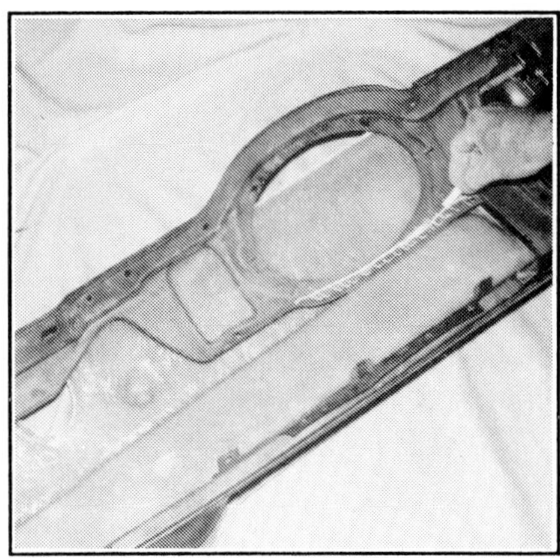

Two cuts must be made to provide clearance for new insert. Chalk mark above speaker opening shows where first cut is to be made.

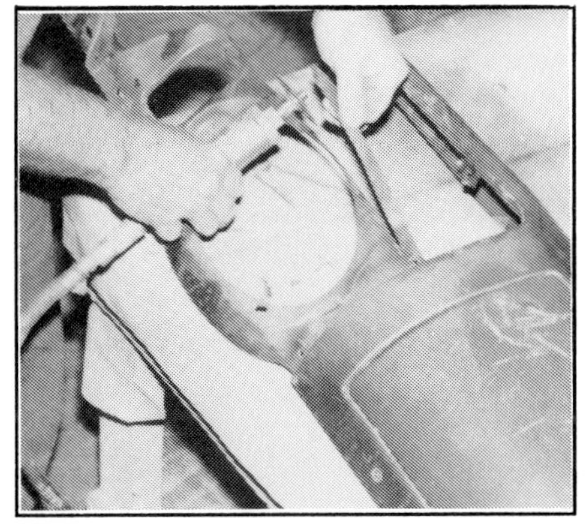

A die-grinder or other suitable sheet metal cutting tool is used to trim opening.

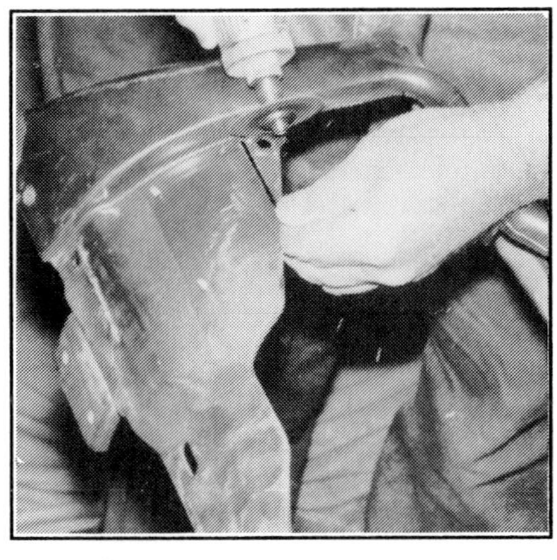

Second cut is made in angled corner at left end of panel opening.

Fiberglass replacement insert was molded with additional material around outside. This permits trimming with a disc sander to fit the opening.

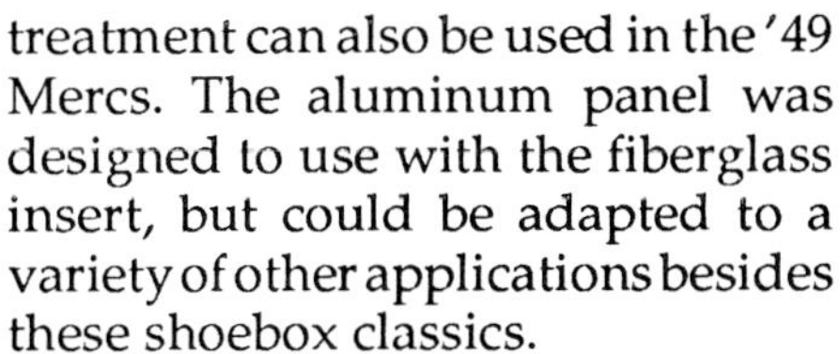

treatment can also be used in the '49 Mercs. The aluminum panel was designed to use with the fiberglass insert, but could be adapted to a variety of other applications besides these shoebox classics.

Updating the control center is actually a pretty simple task, utilizing these new components. Both dash panels are available from Dick Dean in Fullerton, California. Gauges of your choice can be selected at any local parts house. Now you can bring your car's nerve center up to date without sacrificing the basic styling so desirable in a custom cruiser.

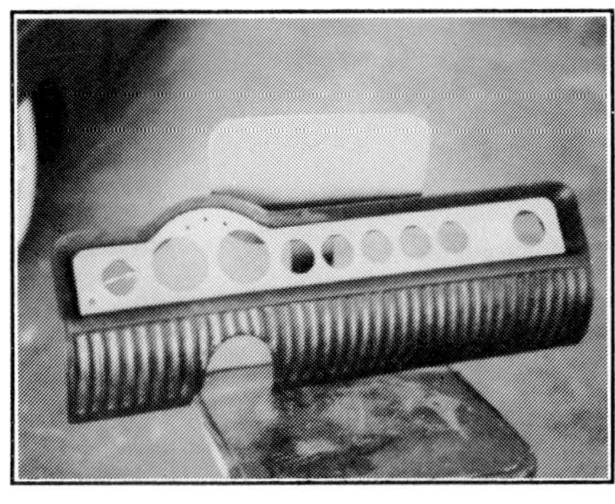

A single large opening or a series of small ones can be cut in replacement insert for clearance of gauges.

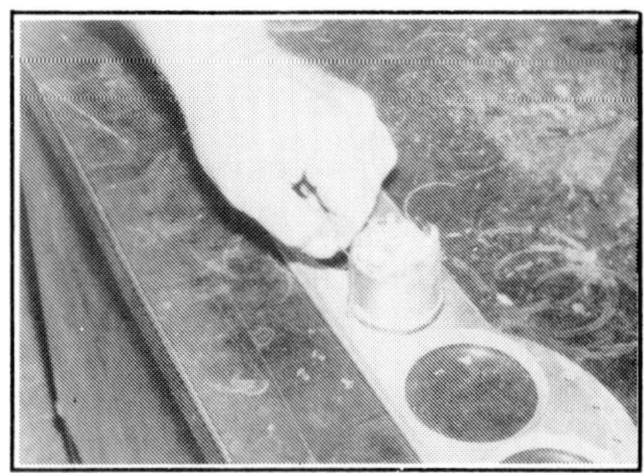

Rear of panel is recessed to accept mounting of gauges from back of panel. Screws inserted in pre-drilled holes fasten gauges in place.

HOT ROD

ORDER HOTLINE 800-782-0539 Mastercard/Visa

Hot Rod History

The legends of hot rodding are raving about this new book! Although the very first hot rod was created when the first automobile was built, it is generally agreed that the modern hot rod sport got its big push during the late 1920s and 1930s. But until recently,any kind of history of hot rodding was ignored. Not now. With over 200 pages literally crammed cover-to-cover, Tex Smith's *Hot Rod History* traces the hobby from Ed Winfield in the Twenties through activity in the early Fifties. First of an on-going series, this volume by Tom Medley will become the most read book in any rodder's library. Featuring such legends as Kong Jackson, Spalding Brothers, Wes Cooper, Bruce Johnston, Bob Stelling, Alex Xydias, Ray Brown, Wally Parks, etc, etc. Photos and facts to delight the most astute historian, but most of all, this is a book that is just plain fun to read

Hot RodHistory softbound................$17.95 $3 shipping
Limited edition hardbound.................$26.50 $3.50 shipping
(Allow 8-12 weeks delivery.)

For Chevy Folks

There has never been a building manual devoted exclusively to Chevy hot rods, particularly anything that deals with the 1954 and older cars (clear back into the 1920's). This book does all that, with over 200 pages crammed with how-to's and premium examples of both backyard and pro shop projects. Everything from bodywork to chassis to suspension and powertrain, this book covers it all. Even material on wooden framework for the body. If you're into Chevy hot rods, then this is a manual you absolutely, positively cannot do without.

How To Build Chevy Hot Rods.....$17.95 plus $3 shipping

For Mopar Nuts

If you're into Chrysler Corporation hot rods and customs, then *How To Build Chrysler/Plymouth/Dodge Hot Rods* is for you. There is no other book like this one, with a concentration on nothing but Mopar products. It is the first of a series, and guaranteed to give you more information than any other source. Packed with how-to's and photos.

How To Build Chrysler/Plymouth/Dodge Hot Rods..........$17.95 $3 shipping

Automotive Art

Every male, from 13 to 85, loves to draw cars. But there hasn't been a real book on automotive art for nearly 40 years. How To Draw Cars cures the problem. Over 200 pages of magic, showing everything anyone needs to know to go from simple doodle to professional drawing. This is the book that dreams are made of, and it is the perfect gift, even for the non-artist. Covers everything from hot rods to dream cars to sports cars and racers. Another absolute must-have for the complete car library.

How To Draw Cars.............$17.95 $3 shipping

Chopping Tops

This is the only book on the subject available, and it is a runaway best seller. Over 200 pages, with step-by-step guidelines for hammering the lid on rods and customs, as well as pickups. Tops fit into 3 basic categories, so what applies to one marque applies to most other makes with the same style top. Covers cars from the Twenties/Thirties through today. How To Chop Tops has become the instant "bible" of rodders and customizers around the world.

How To Chop Tops..........$17.95 $3 shipping

For Fat Ford Fanatics

The 1935-48 Ford and Mercury is firmly established as the current hot rod of selection. But there is precious little building information available on these popular (and readily available) members of the legendary Ford family. Everything is included in this book, with hundreds of how-to photos and drawings...even frame drawings with measurements. Nothing but Fat Ford products in this 200 plus page manual, there is also a special section devoted to using Ford OHV engines, so that the end product can be a Ford front to rear. This is the most thorough, most complete Fat Ford builder's guide ever published.

How To Build FAT FORDS......$17.95 plus $3 shipping

LIBRARY

Shipped UPS for quick delivery, add $3 postage for each book ordered ($4 foreign)

Basic Customizing

The number 1 book in a special series for car enthusiasts, *How To Build Custom Cars* has quickly become the modern "bible" for the hobby, appealing to the home builder and professional alike. Covers everything from chassis swaps and frame clips to fender flares and sectioning, grille swaps, upholstery...over 200 pages jam packed with nothing but custom car building information. The only total custom book on the market.

How To Build Custom Cars..................$17.95
$3 shipping

Wiring Rods/Customs/ Kit Cars

Automotive electrics can be a huge problem, even for the experts. *How To Do Electrical Systems* takes the mystery out of the process with easy to read diagrams. Anyone can use the book to wire and troubleshoot practically any special built car. Wiring schematics for myriad accessories common to the contemporary rod or custom. The troubleshooting section makes this an invaluable "on board" companion for every builder.

How To Do Electrical Systems................$13.00
$3 shipping

Basic Hot Rodding

How To Build Real Hot Rods is the number 1 book in this series, with an emphasis on low cost how-to's aimed at the home builder. Crammed with information that beginners need and advanced rodders use as reference. Everything is included, from frames to suspension to bodywork and upholstery. This is a "must have" book for any rod enthusiast's library, the foundation for more advanced building.

How To Build Real Hot Rods.........$15.00
$3 shipping

The Camaro Supplement

CAMARO Untold Secrets 1967-1969. This is the book that Camaro enthusiasts have all wanted. Camaro expert Wayne Guinn has gathered the information from his years of doing performance research on America's next major hot rod with Chevrolet and Penske Racing engineers. The result is a book that fills in all the questions about factory and race team development of Camaro performance parts.This book from the insiders clears up all the myths and half-truths about the car destined to be one of the most popular hot rods of any era.

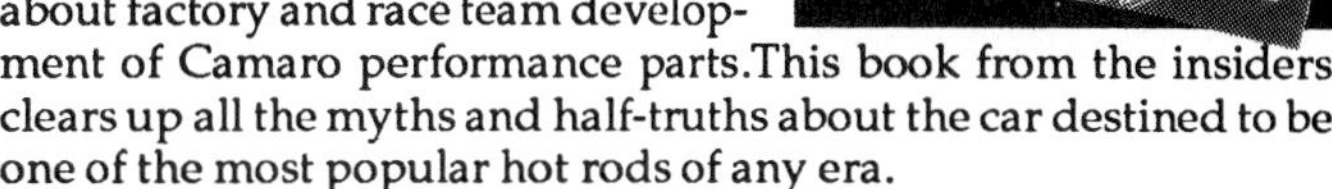

CAMARO Untold Secrets 1967-1969. . . .$17.95
$3 shipping

Available Soon. . . .

Ratfink and Me. The auto biography of Ed "Big Daddy" Roth and his loveable Ratfink character. Roth captured the imagination of a generation of young people during the 1960's, now that excitement has returned to captivate a new group of young (as well as rekindle memories of the older folks). Crammed with pictures.

Inside Hot Rodding. The book that all hot rodding has been waiting for. By legendary hot rodder LeRoi Tex Smith, this is the "word", from someone who has lived the hobby/sport and associated with all the names. The book that tells you what you didn't know who to ask.

How To Build Hot Rod Pickups. Author Jim Clark has been involved with hot rodding trucks since the 1950's and he puts it all down in this one-of-a-kind book. Chassis, bodies, engines . . . everything is here to help anyone who wants to have a hot rod hauler.

How To Build Modern Hot Rods. This is the second in our series of how-to books on basic hot rod construction. The first book *(How To Build Real Hot Rods)* is aimed at the person who wants to do most of the work. This book is for the same person, but concentrates on building/using contemporary hot rod techniques. It is also for the person who wants to buy all the mail-order components to bolt together a great street runner.

TEX SMITH PUBLISHING • 29 EAST WALLACE, PO BOX 726, DRIGGS, ID 83422

by Scott Smith

SAVING A GRILLE

Elsewhere in this book about Shoebox Fords/ Mercs, there is an article about Hugo The Ugh, a 1950 Ford two-door that I once owned. Hugo was in really great condition, but several years of disuse in the California smog had laid a toll on the trim. Especially bad was the grille assembly.

At first glance, it seemed I would need to completely replate the various grille pieces. But after close inspection, it was found that other than a dent in the upper right surround, the original plating was in excellent condition. It was covered with a hazy scum that could be removed. If I took my time, and worked carefully, I could even remove the dent without destroying the plating.

If your boxcar came from an area of low humidity, chances are good that the trim will be in good condition. The best cleaning and inspection can be done only when the pieces are completely disassembled.

Working with ordinary kitchen cleaning soaps, much of the scum could be removed from the grille. Both 409 and Simple Green seem to do an excellent job of cutting through any build-up.

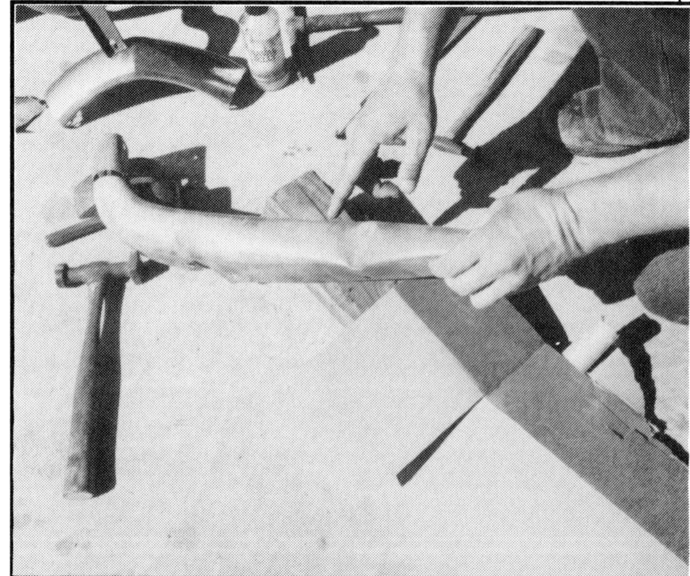

This dent in the grille surround looks much worse than it actually is. Fortunately, the surface was not scratched in the process.

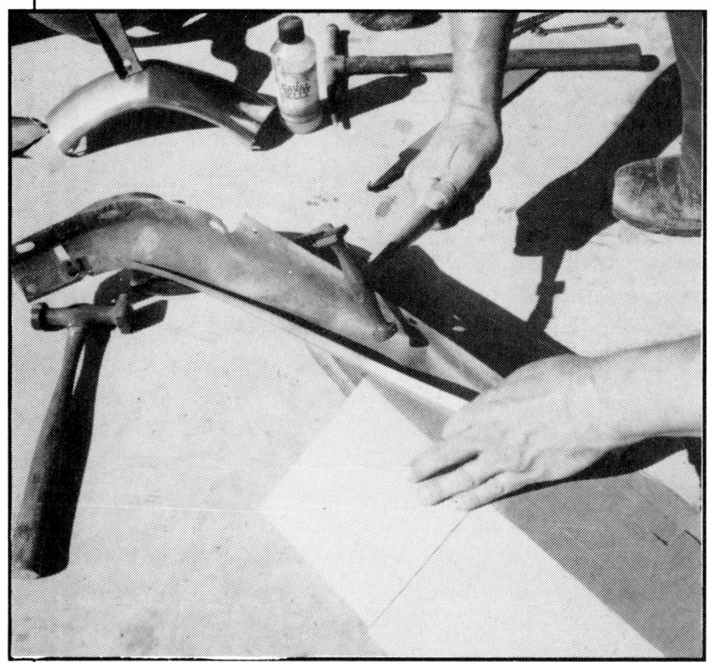

The grille piece was straightened by working from the backside, using a piece of soft pine wood as a support. Use light hammer strokes to avoid stretching the metal, work patiently. By working from the backside, the shiny surface will not be marred by the hammer. Finish by covering a piece of flat iron with a soft cloth (as the work support), then very lightly work the metal with a hammer. If the piece is to be replated, the front side can be smoothed with ultra-fine file and sandpaper.

An experiment with Naval Jelly was tried, and it worked. The jelly is supposed to be for rust, but it cuts scum very well. The secret is not to rub the surface too hard with double-O steel wool. Once the greatest part of scum build-up was removed, then very mild enamel paint rubbing compound was thinned with water and used as a polish.

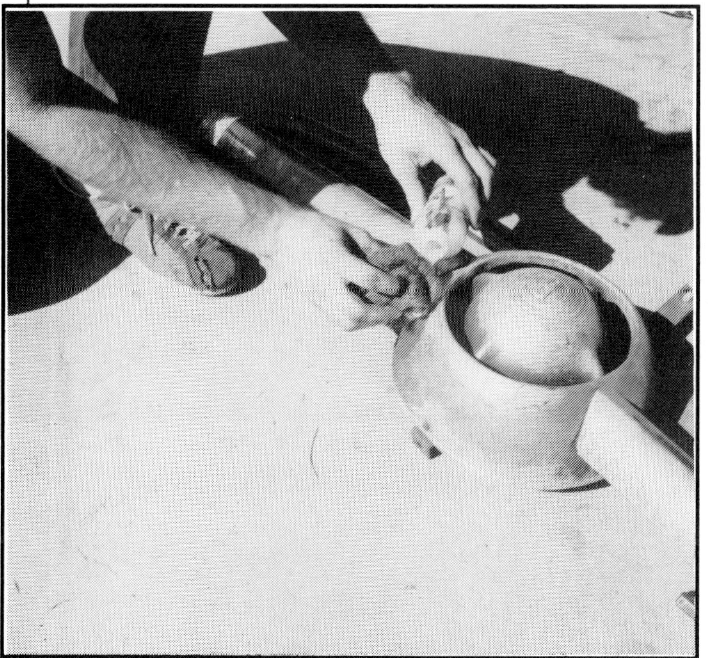

Plating is really a very thin coating on metal, and it can wear away through the years. Be very careful when doing an initial cleaning, and even a worn surface can be saved for some time before replating is necessary. If the trim happens to be stainless, it can be straightened, filed, sanded, and buffed to a better than original sheen. Once this grille was cleaned, it was given a heavy coating of carnauba wax.

More Headroom In A CHOPPED MERC

Pat Palin, of Prineville, Oregon found a way to increase the rear-seat headroom in his chopped Merc. The whole issue of improving the headroom came up when it became necessary to haul a couple of passengers for some great distance in the rear seat of his sled, and there was a certain amount of suffering involved. When the ride was over, Pat went out and started disassembling the Merc interior and making some changes.

The front split-back bench seat is out of a '74 Ford Torino and the rear seat came from the back of a 4-door '70 Oldsmobile. The Olds 4-door seat is wider than the stock Merc rear bench (no armrests) and is thinner but more comfortable. When Pat installed the seat, he pulled the bottom forward about 6 inches and used a new seat riser ahead of the stock one. This new riser measures 6 inches from floor to top, whereas the stock riser was only 4 inches tall. This raises the front of the seat and tilts it back somewhat like a recliner. Oddly enough, even though the front of the seat is raised a couple of inches, the recliner-style position results in improved headroom for rear-seat passengers.

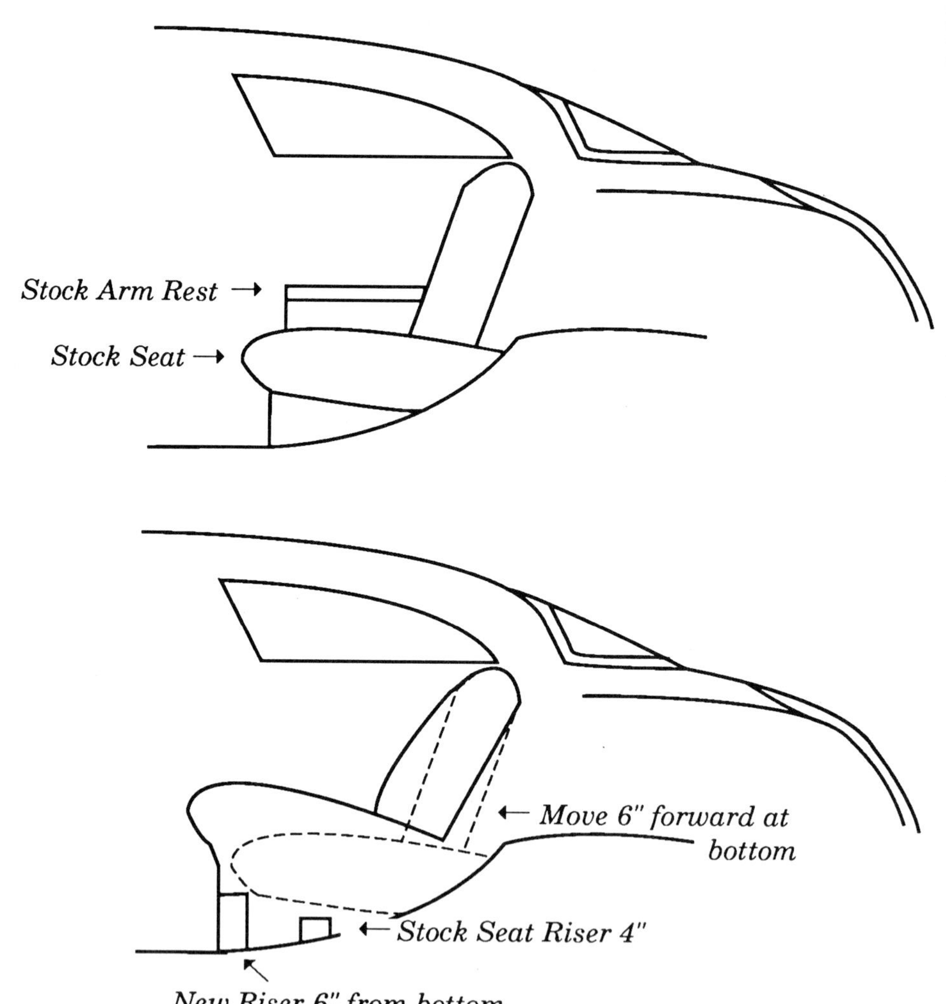

"BRIZ"™ alloy "RIBBED BUMPERS" for "SHOEBOX" Fords & Mercs

'49 / '50 Mercs

'49 thru '51 Fords

Special Applications

Manufactured from high strength 6063 aluminum alloy with a unique deeply ribbed face, contoured to fit each application and polished to a HIGH luster. Stocked for most '48 thru '53 Fords, Mercs, Chevs & Plymouths. We also custom manufacture bumpers in any length and/or roll to your template (usually at no extra cost) - call for information on your particular application **$159ea**

'49 Chev Style License Plate Guard . . . . **$99 ea**

Bumpers by 

Call us with your specific needs! — **206-573-8628**

8002 NE Highway 99, #260 - Vancouver, Washington 98665

by Calvin Mauldin

ROUNDING CORNERS

A customizing trick that is both simple and effective in the appearance department is rounding corners. One doesn't need the skills of a journeyman metal fabricator to whip this one out, though a bit of experience with a gas welder is helpful. Using Jim Baker's chopped '46 Merc coupe as a model, we followed Danny Webster, of Precision Paint and Body, through this "how-to" sequence. With a few minor changes in technique, this will work on doors and hood corners as well. All this custom stuff can be done with simple hand tools, other than a gas or MIG welder, which you might borrow from a friend.

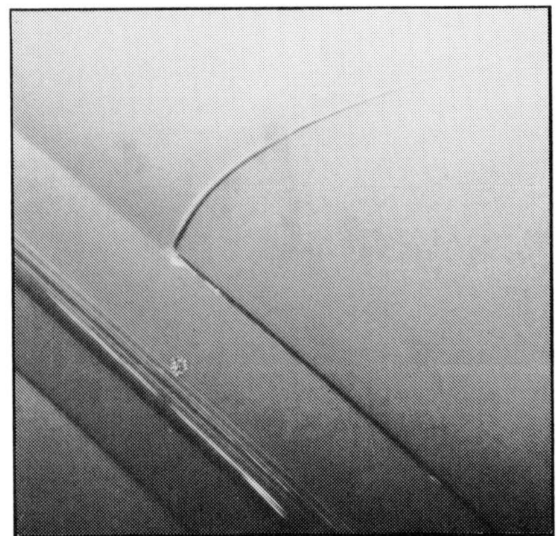

On basically round cars, such as Fords and Mercs of '49-'54 vintage, square trunk corners don't really fit the theme. This is especially true if the top is chopped, lending a more aerodynamic appearance to the car. Rounded corners would look better.

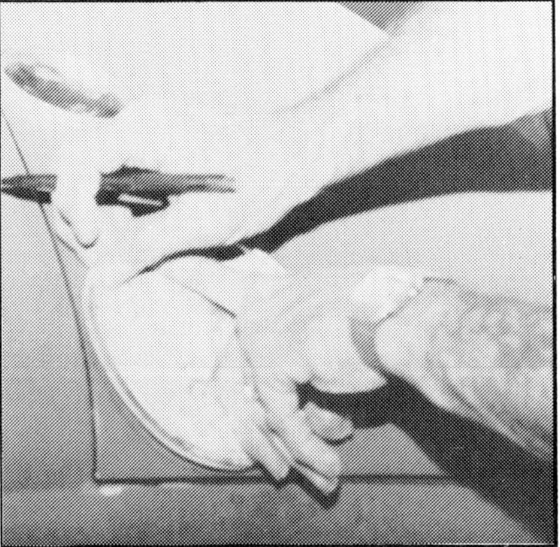

Old paint can lids work well in determining the proper radius for rounded trunk corners. Poster board, properly laid out, can also be used for a template.

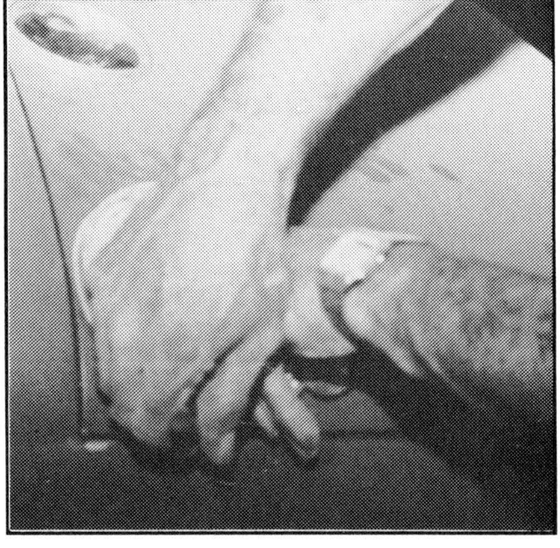

With the lid bent to conform to the trunk's shape, a pencil line is traced onto the surface.

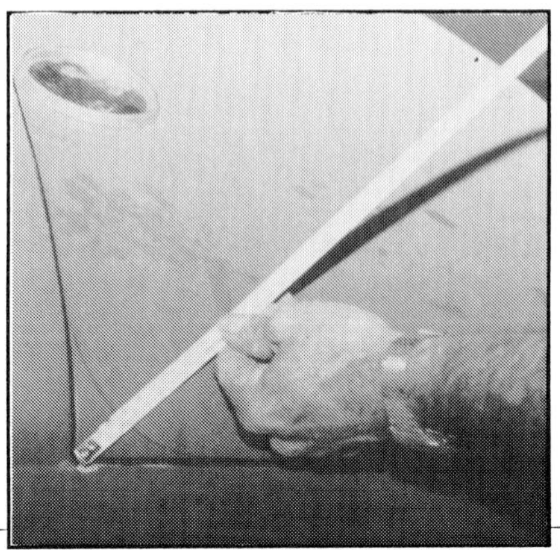

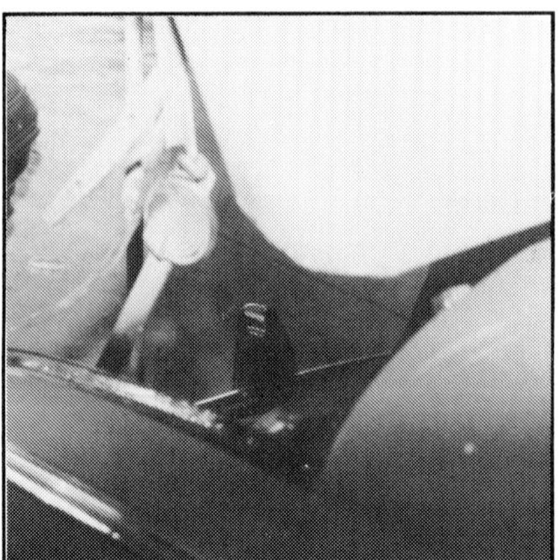

Far Left- Careful measurement should be taken to ensure that the angles on the left and right sides match perfectly.

Danny got pooped out trying to cut through the sheetmetal with a hacksaw, and used a plasma cutter instead. Remember, quick is not always better, because all the time that was saved was lost grinding the radius back smooth. The neater the cut, the less manual labor later.

Be sure to save all the pieces of metal you cut off, because they will be needed to finish the project.

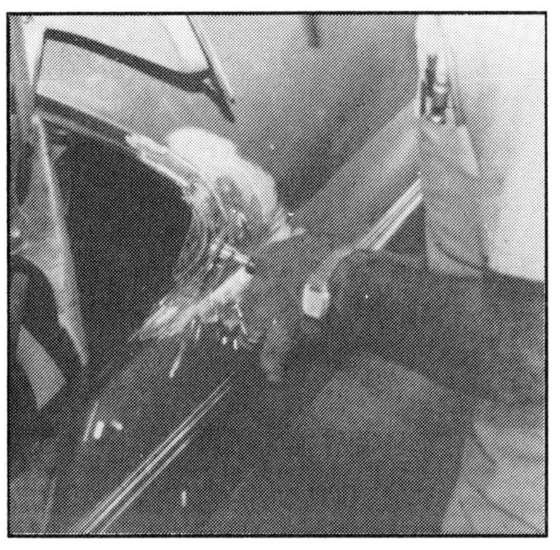

Here Danny does the final grinding. The more time you spend here, the more dividends it will pay in the end.

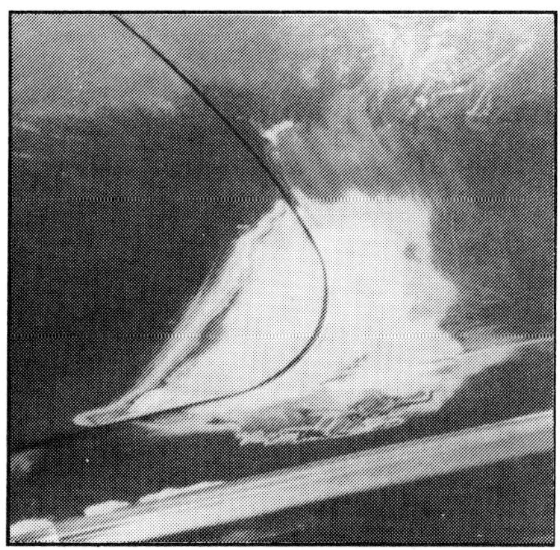

After the block sanding, only a thin film of putty was left. Not bad.

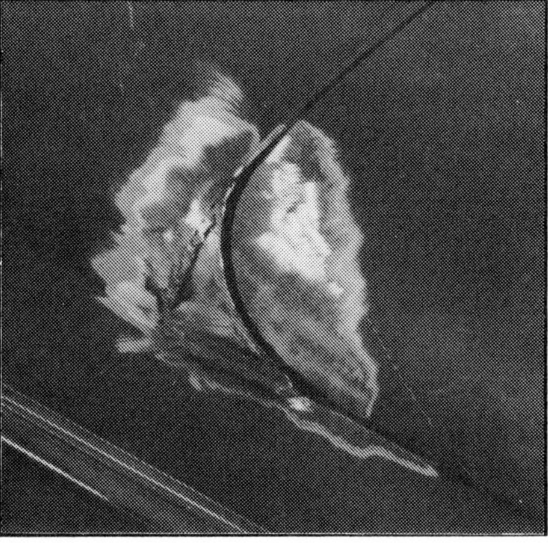

This is where the leftover pieces are utilized as fill-ins. The right and left sides have now been roughed in and ground down.

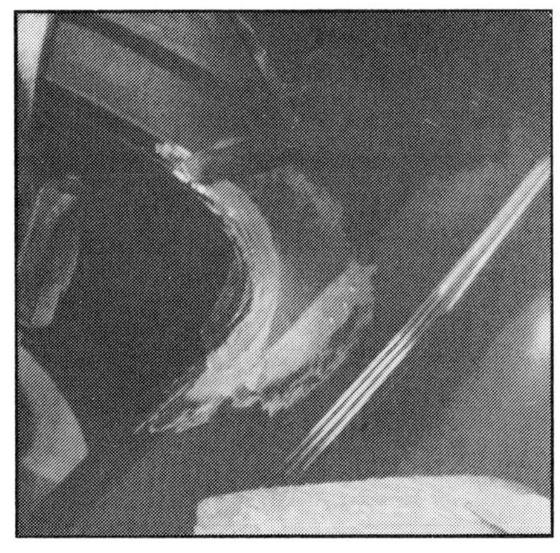

It may look like our man is getting carried away with "the body man in a can" but that's his technique. Put it on thick, then bring it down.

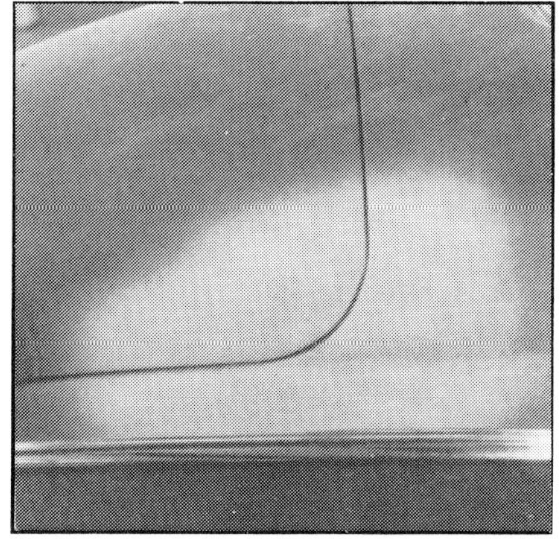

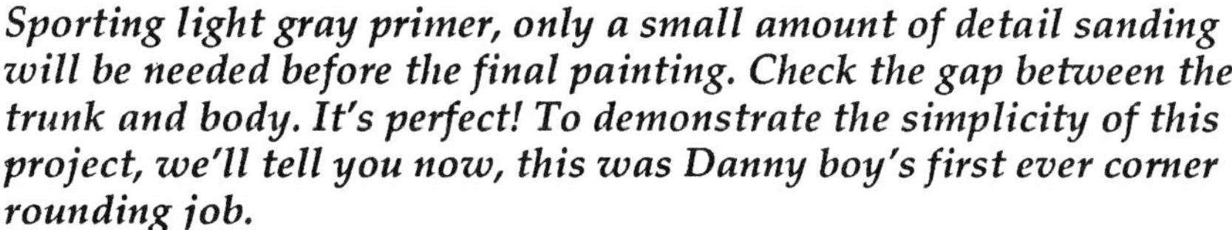

Sporting light gray primer, only a small amount of detail sanding will be needed before the final painting. Check the gap between the trunk and body. It's perfect! To demonstrate the simplicity of this project, we'll tell you now, this was Danny boy's first ever corner rounding job.

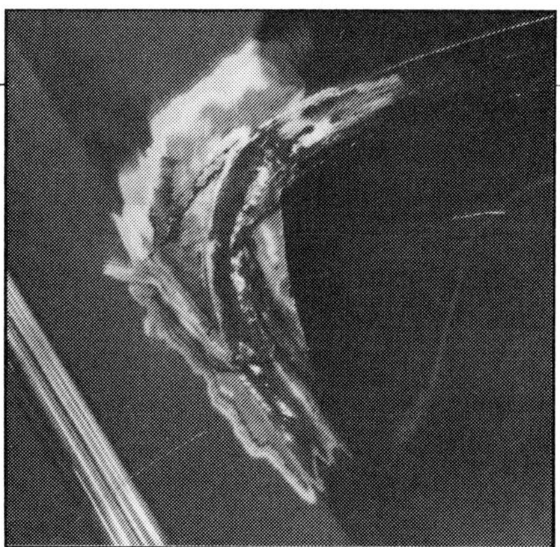

Inside the trunk has to be reworked to conform to the radius. The weather stripping needs to seat properly here to prevent water and dust leaks. Additional metal may be needed to fabricate the shoulder. If you're just getting into customizing, it's wise to keep an old junk hood or fender around for sheetmetal scrounging.

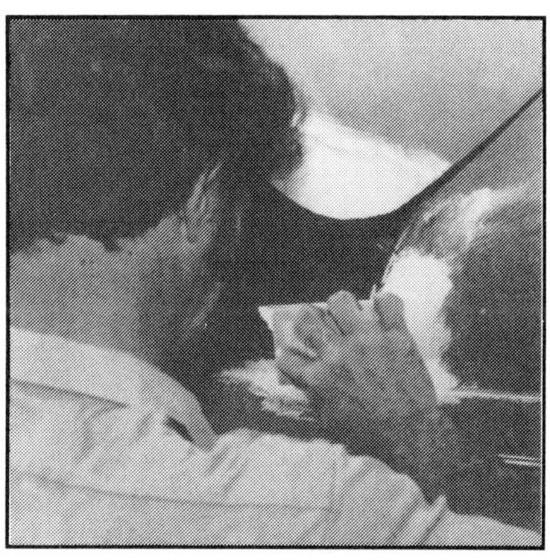

Blocking it to the right contour. The inside edges can't be blocked. Here you'll have to improvise when sanding the surface that meets the weather stripping. Danny wrapped 80-grit sandpaper around his index finger and gave it a go. It worked, but then he couldn't play the banjo for a couple of days.

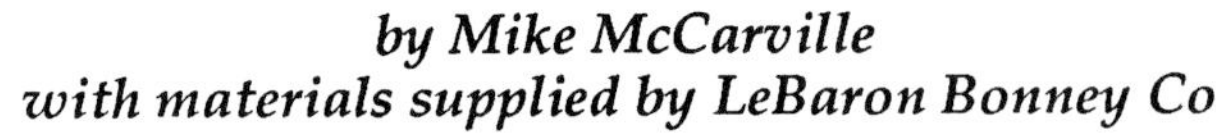
by Mike McCarville
with materials supplied by LeBaron Bonney Co.

REUPHOLSTERY

Upholstering seats with a prepared upholstery kit is a matter of taking your time and having the right tools on hand. As you will see in reading the following LeBaron Bonney Upholstery Kit Installation Instructions, it's not too difficult an undertaking.

These instructions were prepared by LeBaron Bonney for its own craftsmen and are based on years of experience installing award-winning Ford interiors. They take you through the interior renewal process for a '51 Ford Victoria in an orderly, organized fashion.

FRONT SEAT DISASSEMBLY

Unscrew and remove the front seat side shields from the sides of the seat frame, then remove the front seat assembly from the car. Remove the cotter pins from the front backrest hinges and remove the backrests. Unbolt and remove the front seat tracks from the bottom of the frame.

UPHOLSTER FRONT BACKREST, PASSENGER SIDE

Remove the old cover, padding and cardboard. Make such repairs to the springs as seem necessary, then sand or tape any sharp or rough places that might cut or abrade the new cover.

Place the backrest unit face up and center the burlap from one piece of the front backrest padding kit on top of the springs so that it hangs over equally on all sides. Hog-ring the burlap to the edge wires about every two or three inches along the sides and at the bottom. At the top, hog-ring it to the zig-zag springs a couple of turns down (toward the center of the unit). Place the foam rubber insulating pad on top of the burlap, and hog-ring it in the same manner as the burlap.

Turn the backrest face down and place the cardboard piece in position over the tubular frame of the rear of the unit. Push the metal tab at the lower outside corner of the frame through the cardboard and fold the tab down. It can be helpful to tape the cardboard to the frame in a few spots to hold the cardboard in place.

Spread cement along the edges of one of the jute pads and the sides of the tubular frame. Lay the jute over the cardboard and fix it to the frame.

Turn the backrest face up and place the cotton padding on top of the foam pad and center it so that an equal amount hangs over on all sides. You may want to hog-ring the cotton in a few places, so it will not shift when the cover is put on.

Turn the backrest upright and slide one of the plastic bags over it from the top. These bags are supplied to help keep the padding in position as you slip the cover on. After the plastic bag is in place, slit open the center third of the top of the bag so that you can remove the bag by pulling it out from the bottom after the cover is in place.

Fold the cover inside out to halfway up its length. Now, with the plastic bag on and the backrest upright, with the front facing you, slip the cover on, pulling down firmly on the folded fabric. Adjust the cover so the seams and welt run straight along the sides and top. Unroll the folded portion of the cover down over the remainder of the backrest. Pull out the plastic bag once the cover is positioned correctly. Form the cover over the springs by pulling it down firmly and hitting the cover with the palm of your hand to smooth and tighten it. The cover should fit snug "and a little more." Adjust the seams and the welting so they are straight and even.

The bottom of the cover can now be closed. Pull the bottom rear edge of the cover down and hook the edge welt under the prongs on the bottom of the frame, first at the center and then at the corners. Then do the remainder of the back evenly, out from the center to the corners. Next, hook the front edge welt

These photos show how the completed front and rear seats look after the reupholstery job. This work is not particularly difficult, and requires only hog-ring pliars.

Front seats are split-back style, so each backrest needs to be upholstered separately.

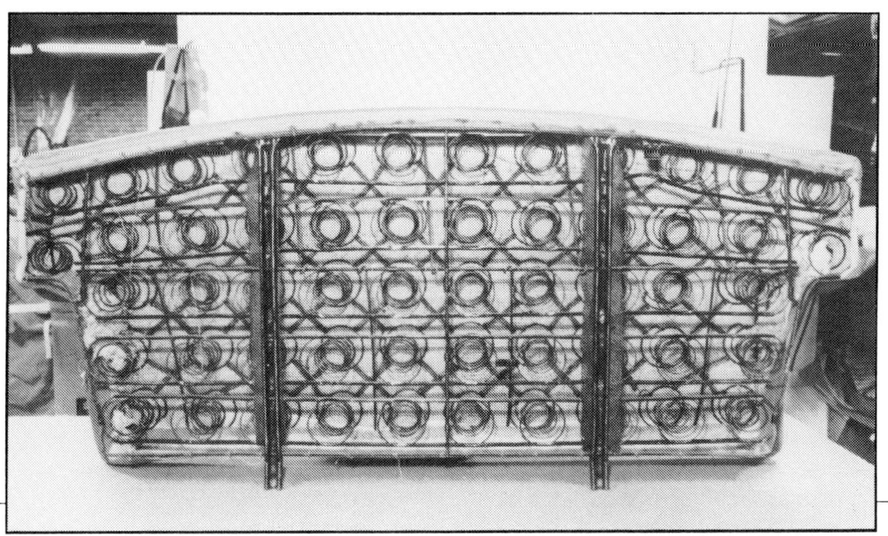

at the bottom of the cover to the prongs in the same manner, or just hog-ring the front edge welt to the rear edge welt. Be sure to leave the backrest stop-bolt head uncovered. To finish the sides of the cover, fold the edges under and fasten with a hog-ring. Repeat procedure for the driver's backrest.

UPHOLSTER FRONT SEAT

Remove the backrest scuff plates and the old cover and padding, and repair the springs as necessary. Sand or tape any places that might damage the new cover.

Place the seat unit face up and center the burlap on top of the springs so it hangs over equally on all sides. Hog-ring the burlap every two or three inches along the rear to the zig-zag springs, and to the upper edge wire along the sides. Turn the unit face down and hog-ring the front of the burlap to the springs three turns down from the front edge, in toward the center of the unit.

Turn the seat face up again. Place the foam pad from the kit, or your original fitted foam pad if it is still in

Rear seat bottom springs are shown here. Springs should be inspected and repaired or replaced, as necessary, before starting the reupholstery job.

good condition, on top of the burlap, and hog-ring all around. Place the cotton padding over the foam so that an equal amount hangs over on the sides. End the cotton flush with the tubular frame at the rear.

Position the cover over the padded springs with the cut-out at the center rear, going around the center hinge of the seat assembly frame. Make sure that the cover is centered side-to-side on the seat frame. Smooth the cover out by pulling and hitting it firmly with the palm of your hand. Again, the cover should fit snug "and a little more." With the cover positioned correctly, and the seams straight and even, turn the unit face down and begin hog-ringing the front edge welt to the edge wire along the front of the frame, first at the center and then at the corners. Next, hog-ring the edge welt stitched to the rear edge (carpet section) of the cover to the zig-zag springs at the center, then the ends. Pull the sides down and hog-ring them to the edge wires at the sides. Cut holes in the sides of the cover to expose the outer hinges for the backrest units. Make sure the cover is aligned correctly, and is smooth and even, then hog-ring every two to three inches all around.

Replace the backrest scuff plates and the seat tracks, then put the backrests onto their hinges and secure them with cotter pins. Bring the center hinge cover flap over the hinge post and fasten it to the rear of the post with the oval-head screw and recessed washer provided. Replace the side shields either now or after the seat assembly has been reinstalled in the car.

UPHOLSTER REAR BACKREST

Unbolt and lift the rear backrest out of the car. Remove the old cover and padding, and prepare the springs and frame as before.

Cover the front face of the springs with burlap, hog-ringing it to the edge wires all around. Then lay the foam rubber insulating pad over the burlap and hog-ring it in the same manner. Lay the cotton padding over the foam, letting an equal amount extend over edges all around.

Center the cover over the padded springs. Smooth the cover by pulling and hitting it with the palm of your hand until the cover is aligned correctly and is smooth, even, snug "and a little more." The welt should run evenly along the front top and side edges. Hog-ring the top edge of the cover to the edge wire at the upper rear of the backrest at the center, then at the corners, then a few places in between. Hog-ring the bottom of the cover in the same manner, and then do the sides. With the cover aligned correctly, and smooth and tight, hog-ring every few inches all around.

UPHOLSTER REAR SEAT

Remove the old cover and padding, and repair springs as needed. Lay the burlap over the springs so an equal amount hangs over on all sides. Hog-ring the burlap to the spring coils one turn down, then lay the foam pad over the burlap and hog-ring in the same manner. Lay the cotton over the foam pad, letting the cotton extend forward from the rear top of the spring unit down the front face of the springs almost to the bottom of the frame.

Center the cover over the padded springs and smooth it out by pulling and hitting it as you did for the other covers. When the cover is aligned correctly and is tight and smooth, hog-ring the front edge welt. Start at the center, then do the corners, and then hog-ring in between. Hog-ring the rear edge welt in the same manner, then hog-ring the sides.

Put the backrest and seat cushion back in the car and check to see that they meet evenly. If they do not, if there are high or low spots on either unit, remove the seat or backrest and adjust the padding as necessary. To do this, just snip off the hog-rings with wire cutters in the affected area, then rearrange the padding and fasten the cover back in place.

Specialty POWER WINDOWS

The Only Company Devoted Exclusively To Manufacturing Power Window Kits For Older Vehicles

---

(912) 994-9248 Technical Information
1-800-634-9801 Factory Sales Desk
1-800-728-3881 West Coast Sales (CA)
FAX (912) 994-3124

Route 2, Goodwyne Road, Forsyth, Georgia 31029

by Jack Chisenhall

AIR CONDITIONING

Keeping Your Cool

As Pete Chapouris once said, "To be cool, you gotta stay cool." As long as our surroundings are somewhere between 70 and 80 degrees Fahrenheit, most of us feel comfortable (that's cool). Below 70 degrees most of us feel cold, above 80 degrees most of us feel hot (and that's not cool). Air conditioning is one way to keep our immediate surroundings within man's comfort zone.

In order to properly select air conditioning system components, it is helpful to understand a few of the basics of how the system works. The following three basic laws of nature are used in an automotive air conditioning system to make it all work. 1. Heat Transfer, 2. Vaporization, and 3. Pressure's effect on vaporization.

1. Heat transfer — Differing temperatures always try to equalize. Heat always travels to cold. The greater the difference in the temperatures, the faster it travels.

2. Vaporization — Heat is released by condensation, and heat is absorbed by evaporation.

3. Pressure's effect on vaporization — Increasing pressure on a liquid raises the boiling point.

These three laws of physics work together in an air conditioning system to remove heat from the passenger compartment and transfer it to the outside. The liquid refrigerant (R-12) is contained in a closed system that has two different pressure areas in it. By moving the refrigerant from the high-pressure side of the system to the low-pressure side, its state is changed from a liquid to a gas. When it changes, it absorbs and gives off great amounts of heat, and the various system components transfer the heat to the outside of the vehicle.

Some important safety precautions must be followed when working with R-12 refrigerant. Because R-12 boils at minus 21.7 degrees Fahrenheit, it is cold enough to cause severe frostbite. Always wear goggles to protect the eyes and gloves to protect the hands. Also, be sure to never expose a can of R-12 to direct sunlight. Even under the pressure normally found in R-12 containers, if heat is added, the refrigerant will boil. This could raise the pressure inside the can to a dangerous level. One last caution — when R-12 is exposed to an open flame or to hot metal, poisonous gas called phosgene is formed. You don't want any of that stuff around — it was used as nerve gas in World War I.

Air conditioning refrigerant travels through the system of components that consist of the compressor, evaporator, condenser, receiver/drier and the various hoses and fittings. These components provide a system in which the R-12 can evaporate and condense to remove heat from the inside of the car and give it off to the outside air stream.

As we discuss the components, keep in mind that all automotive air conditioners work basically the same way. Three distinct types of systems are in common use now. The difference is in the control of the flow of the refrigerant. One type is a cycling clutch expansion valve system (CCEV). The second is a cycling clutch fixed orifice system (CCOT). The third is the pressure regulating expansion valve system.

The first two types (CCEV and CCOT) control the temperature by cycling the compressor to stop refrigerant flow and avoid freezing of the evaporator. The third type controls the pressure with the evaporator, thus controlling its temperature and avoiding freeze-up.

Aftermarket manufacturers use the CCEV system with rare exceptions, primarily because it remains the most efficient in operation and works well with the type of evaporator coils that are most often used.

The CCOT system is rapidly becoming the most common original equipment (Ford and GM) type system. It is inexpensive to produce, its fixed-orifice metering device is molded plastic with no moving parts, compared to the expansion valve's relatively costly construction. The CCOT system floods the evaporator with refrigerant, because it does not meter according to evaporator demands. This flooded condition requires an accumulator, to prevent unused liquid refrigerant from returning to the compressor, and uses increased horsepower to drive the system. Head pressure is higher as a result of this flooded evaporator condition.

The CCEV system is the most common in hot rods and will likely remain so for a while.

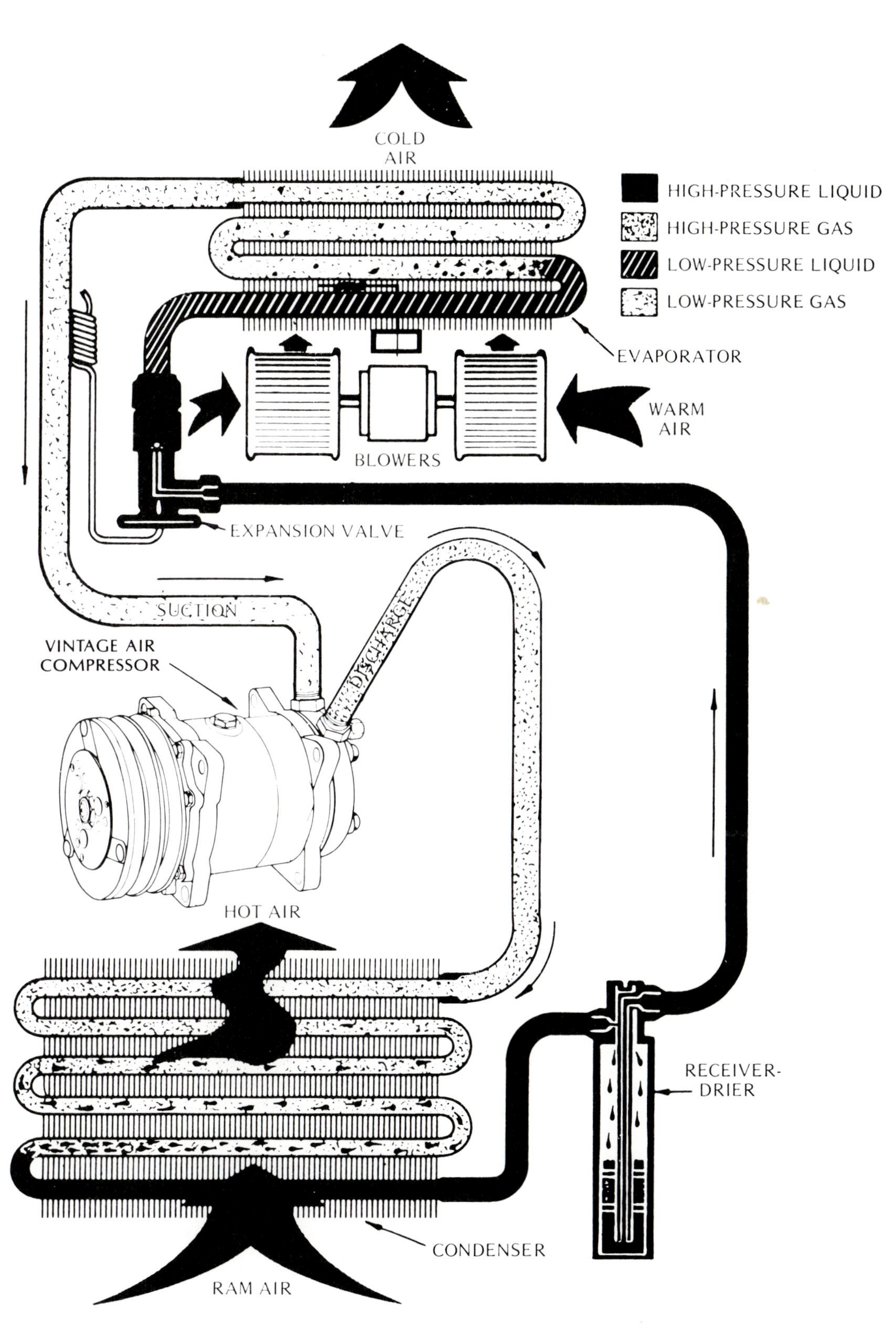

This schematic illustrates how the refrigerant travels through a typical air conditioner system.

CONDENSER

The condenser is the most misunderstood, mismatched and misplaced component in street rod air conditioning. Space, air-flow and air temperature will affect condenser selection. As a general rule of thumb regarding two-row 5/16" or 3/8" copper tube and aluminum fin condensers, surface area should be a minimum of 210 square inches. These condensers are normally 1.25" thick, giving about 262 cubic inches of mass. This basic minimum assumes a few other variables.

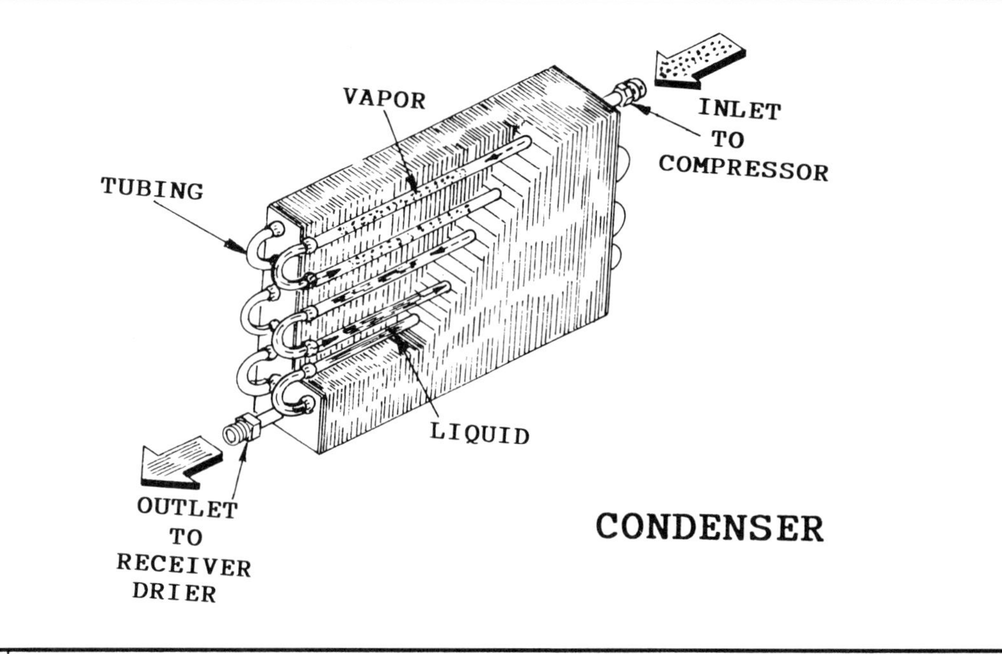

Vaporous refrigerant condenses as it is cooled by the air passing over the condenser. The condensing refrigerant gives off heat that was absorbed when the refrigerant evaporated inside the passenger compartment evaporator.

1. Ten or twelve fins per inch is ideal, letting enough air-flow through the condenser.
2. Air-flow should be good at low driving speeds (approximately 1500 to 2000 cfm).
3. The evaporator is about 210 cubic inches of mass. The condenser should be about 20 percent greater in mass than the evaporator. Again, this is a rule of thumb with many variables.
4. The condenser should be mounted so that the tubes run horizontally. If not, the oil that flows with the refrigerant will settle in the loops at the bottom of the coil. This restriction will cause a corresponding rise in system pressure and refrigerant temperature, reducing the ability of the condenser to dissipate heat.
5. Compressor size is approximately nine cubic inches per revolution. This has proven more than adequate for any single-evaporator system. I would always lean toward a smaller compressor than a larger one. If you have a special or larger-than-typical application and want to increase the refrigerant capacity, increase condenser size not compressor size.

The temperature of the air flowing across the condenser should be 100 degrees Fahrenheit or less. The greater the difference between the air and refrigerant temperatures, the greater the heat transfer. That the air temperature is below 100 degrees Fahrenheit is more critical to the condenser than the engine radiator, because refrigerant boils at a much lower temperature than water (coolant), and begins expanding and creating pressure at a very rapid rate above 100 degrees F.

The condenser should always be mounted in front of the radiator. Here it uses less plumbing, takes advantage of the engine fan, and is in the best place for the outside air stream which is the coolest air available. If you use two condensers to do the job, the last one in the series should be in the coolest air stream. Leave 1/8" to 1/2" distance between the radiator and the condenser. That's enough to prevent metal-to-metal heat transfer, but close enough to ensure air-flow. Condensers can be mounted anywhere, but ignoring the basics will leave you with a system that doesn't work.

Condenser construction falls into two basic types. They are either serpentine — a continuous flat tube with multiple passages that run back and forth with fin material between the tubes to form a flat slab (serpentine condensers are generally extruded aluminum with fittings welded on each end of the tube); or they are "tube and fin" type — round tubes (usually copper) running back and forth through flat fins (often aluminum) to form a flat slab. I generally prefer copper-tube-and-aluminum-fin condensers because they are easily modified with a torch and silver solder.

A larger condenser is almost always better. We are trying to remove heat. Heat and pressure within the A/C system are directly related. The more heat we remove, the lower the pressure within the high side of the air conditioner, and the better we've done the job. This lower pressure/temperature also requires less horsepower to turn the compressor and causes less vibration, which is the most common cause of compressor mounting bracket breakage.

COMPRESSOR

The compressor must be sized right for the system, and it is important that it looks good under the hood. We have found that with at least a basic minimum condenser and a typical evaporator of about 220 cubic inch mass, an 8.5- to 9-cubic inch per revolution compressor capacity is ideal. If you are short on condenser and cannot simply increase its size, you might opt for less compressor capacity. Generally speaking, if the compressor is oversized in relation to the condenser, it will create excess pressure and heat, placing extra load on the engine and leading to premature compressor and mount failure. Lack of compressor capacity will normally result in reduced performance, but you would probably only notice it at a stoplight or in traffic. This might be a blessing, however, as you would also be adding less heat to the radiator. So, if it sounds like I'm saying to downsize the compressor, maybe I am.

The Sanden compressor is compact, smooth and attractive. The compactness of a Sanden unit makes it easy to install in a cramped engine compartment. The small 505 unit is available for very crowded compartments.

The York/Techumseh-style compressor was once the most common, but the Sanden (previously Sankyo) has closed the gap with its more contemporary radial design. Sanden/Sankyo compressor is about five inches around with a length of 8-1/2 inches for the 508 model and 6-3/4 inches on the 505 model. Model 708 is a seven-cylinder unit totaling eight cubic inches per revolution, and it weighs 18 percent less than its predecessor 508 (which has five cylinders and eight cubic inches per revolution). More cylinders means greater smoothness while increasing volumetric efficiency by seven to ten percent at 800 rpm. Many other radial designs are available including Frigidaire, Nippondenso, and Diesel Kieki (licensed by York in the U.S.). Let's take a look at each.

Frigidaire — The time-honored A-6 is a real dinosaur. Steer clear unless you are building a limousine with room for a dinosaur under the hood. Its 12+ cubic inches per revolution is more compressor than you would normally need or want! The R-4 is the short, fat compressor found on the later '70s and early '80s GM cars. It has 10 cubic inch per revolution displacement via four cylinders radiating from its crankshaft. Smaller and lighter than the A-6, it is still larger than I like in capacity.

Diesel Kieki — This is a clone of the Sanden/Sankyo and will mount with commonly available brackets. However, I have found that they are not nearly as durable as the slightly more expensive Sanden.

Nippondenso—This compressor is original equipment on the new Ford and Chrysler cars. Many versions exist and most are sized at just over nine cubic inches per revolution. The fittings connect in the middle of the compressor, making it difficult to fit a pleasing hose arrangement. Although these compressors have a little more capacity than I would want on most early street rods, they may be well suited for later cars with larger condenser area and cabs. Brackets, however, are limited to original equipment.

York/Techumseh — Radial compressors should be, in my opinion, the only style compressor you consider. The old style York/Techumseh reciprocating two-cylinder compressor is simply outdated and no bargain at any cost. It can be identified by its square shape, closely resembling a lawn mower engine. Its displacement is on a plate riveted to the front of the compressor in a space marked "Pt" or "Part" and it will say something like R-209. The 209 tells us it is a two-cylinder compressor with nine cubic inches of displacement per revolution. The three compressors in this line are the 210 (10.3 cubic inches), 209 (8.7 cubic inches), and the 206 (6.11 cubic inches).

The major drawback of a reciprocating compressor is that, because it has only two cylinders, the peak torque loads are very high on each stroke, causing excessive vibration. This makes for lots of noise, broken parts and drivebelt flop.

Another consideration for compressor selection is the way it fits into the available space and how it mounts to the engine.

COMPRESSOR BRACKETS

Brackets that will fit most applications are available for both the York and Sanden type compressors. The radial compressors are easiest to mount because of their alternator-style construction. They do not shake as much, so they require less bracketry. Generally, building brackets for air conditioning compressors follows the same principles as any other steel bracket design and fabrication — the simpler the better. A few things, however, are peculiar to this type of bracket.

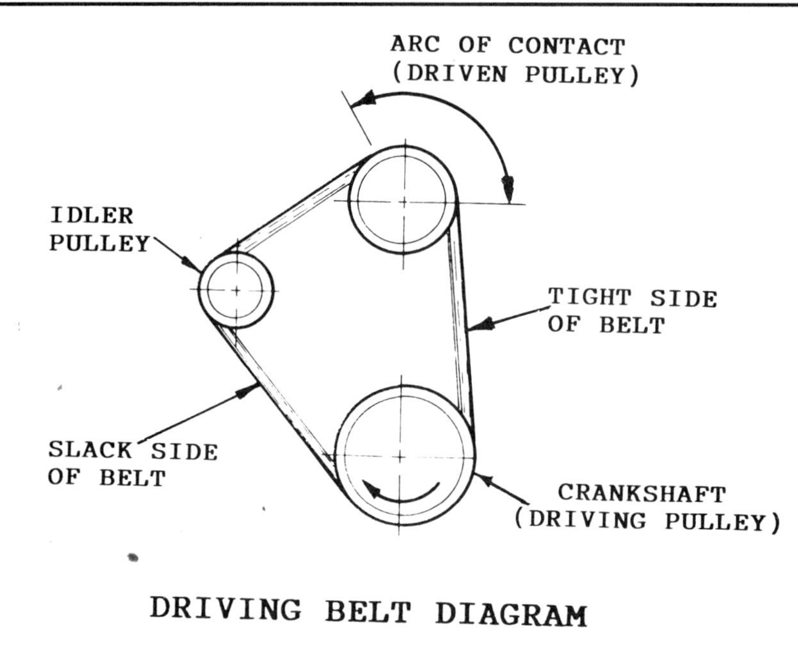

When installing the compressor, make sure that the drive belt makes contact on at least 33% of all driving pulleys. The drive belt has a tight side and a slack side. When using an idler pulley to adjust belt tension, put it near the center of the slack side of the belt.

There must be a way to adjust the tension of the driving belt. Sliding the compressor, or using an idler pulley mounted on an adjacent eccentric are the simplest methods. I don't like to use idler pulleys because they wear out rapidly and are excess hardware. If you elect to use an idler pulley, however, (and sometimes you must), try to put it on the slack run of the driving belt. A driving belt always has a tight side and a slack side. By doing this, the bearing life of the idler is increased by putting less load on it, and belt flop is reduced. Make sure there is enough belt contacting all driving or driven pulleys. This area of contact should be no less than 33 percent of the pulley's circumference.

Brackets are commercially available for most common applications like small-block Chevy, Ford and Chrysler engines. If you need special brackets, universal mounting plates are available for building your own for both York/Techumseh and Sanden compressors.

HOSE AND FITTINGS

Hose and fittings are most commonly used to plumb an air conditioning system. Because of the exacting duties that mobile fluid lines and fittings have to perform, a few industrial groups have set some standards. The three that apply here are Society of Automotive Engineers (SAE), Joint Industrial Council (JIC), and International Mobile Air Conditioning Assn., Inc. (IMACA). Automotive tubing and fittings are built to SAE specifications. Within the aircraft industry, manufacturers adhere to JIC specs. The aircraft parts would not apply to us except that someone decided a long time ago that it was neat to use airplane parts on a hot rod. Many of us do just that, so a little knowledge of these parts can't hurt, especially if you might be trying to mix JIC parts with SAE components. The IMACA standards that apply are the test standards that the parts should meet.

SAE/IMACA Hose and Fittings — Automotive fitting manufacturers make an incredible number of different fittings for almost any situation. Knowing where to find that fitting is the trick. The simplest way to plumb your air conditioner is to use these fittings, available at auto air conditioning supply outlets.

Vintage Air (10305 IH 35 N, San Antonio, Texas 78233; 512/654-7171) has come up with a line system of stainless steel tubing and SAE/neoprene O-ring fittings that will complement the best-detailed car. A small amount of flexible refrigerant line isolates the engine's movement.

AN/JIC stainless braided line and fittings — This type of hose and fittings has found fairly widespread use on street rods. It is visually appealing, however these fittings can be a real pain when it comes to making them work. JIC flare fittings have 37-degree seats, while automotive air conditioning components are designed with 45-degree SAE seats on their fittings. When you place a 37-degree flare fitting on a 45-degree seat, it cuts a ridge in it. This may seal well once, or even a few times, but eventually the seat will become defective. But that isn't the real problem. The problem is really in the hose and its connection to the fitting. This hose is designed for hydraulics, not refrigerant-12. The manufacturers' engineers flatly state that the Teflon or neoprene hose liners will not contain R-12. Further, the connection between the hose and the fitting is not suitable for containing R-12. Nonetheless, street rodders continue to use it. Just be aware of what goes with the territory.

DO'S AND DON'TS OF A/C PLUMBING

DO push the hose all the way on the barb of a fitting.

DO use rubber hose inside the cab on the suction line, or wrap any metal line or fittings inside the cab with insulating tape.

DO try to tie the suction line (cold line from the evaporator to the compressor) to the liquid line (from the condenser to the drier then evaporator) so that heat transfer takes place from the hot liquid line to the cold suction line. This helps further remove BTUs from the liquid refrigerant and further vaporize the refrigerant going back to the compressor. Compressors are designed to pump vapors only; liquid droplets are hard on the delicate reed valves that separate the high side and the low side.

DO NOT allow the hose to rub against sharp points.

DO NOT allow the discharge line (from the compressor into the condenser) to contact the liquid line (from the condenser to the drier and then evaporator).

DRIER

Moisture must be removed effectively from the air conditioner. This is the greatest problem in any refrigeration system. Moisture combines with the metals in the system to produce oxides, primarily iron hydroxide and aluminum hydroxide. It also combines with R-12 to produce acids, notable carbonic acid, hydrochloric acid and hydrofluoric acid. These acids attack the base metals within the system, producing a substance like a combination of dust and sand.

The drying function of the receiver/drier is a simple process. The drying agent (desiccant) is held inside a felt bag placed within the receiver tank. When the refrigerant flows across and through the desiccant, moisture is captured.

The drier also acts as a surge tank to take the sudden surges when the engine accelerates quickly. This keeps the pressures throughout the system fairly constant.

The sight glass is a window to the refrigerant leaving the drier. Pure liquid refrigerant is clear. When there is no liquid seal in the bottom of the receiver tank, the refrigerant will draw gas bubbles into it. These will make the refrigerant foamy. Discoloration means a contaminated system. This is a guide only. If you are not sure, ask a qualified A/C technician.

DO mount the drier according to the manufacturer's specifications to ensure a liquid seal.

DO mount the drier in the coolest spot you can — in the air stream if possible.

DO install the drier where you can see the sight glass.

DO replace the filter/drier any time the receiver is exposed to the atmosphere for any period of time, or whenever you perform any maintenance on the system.

DON'T neglect drier maintenance.

DON'T allow excessive heat around the drier.

DON'T lay a vertical-style drier on its side.

EVAPORATOR

From the drier, the refrigerant travels to the coolest component in the system — the evaporator. This is the part that draws air in the cabin, circulates it through the evaporator coil to remove heat and moisture, then returns it to the cab as cool, dry air. A typical evaporator has a coil, an expansion valve, a housing, a blower and air outlets.

When selecting an evaporator, the two considerations have to be, 1. Will it fit in the car, and 2. Will it cool the car adequately? Choosing the appearance you want, there are two basic choices: under-dash and in-dash units.

Under-dash units work well in some applications. The contemporary under-dash unit is the slim-line type. Its stylized design has a long, thin front bezel containing multiple air outlets. The direct air delivery of this type unit is very effective and it does not interfere with glove box, radio or other in-dash accessories.

With an in-dash unit, the big questions are, 1. Will it fit in the dash, and 2. What will you have to give up to get it there? In-dash units with molded ducts from the evaporator coil are found in the higher-performance applications such as vans and better-quality aftermarket manufacturers. This allows direct, large volume air delivery from a concealed unit.

Selecting an evaporator involves consideration of the size you are trying to cool. If you have a coupe, you can obviously use a smaller unit, a sedan would require more capacity, and so on.

Fitting the unit to your car can be very critical if you are ordering by mail or fitting a non-standard vehicle. The best way is to start by looking at the space behind the dash and trying to make a rough sketch of the area between the firewall and the dash and then over to the kickpanel. Note radios, instruments and other difficult-to-remove items. When talking with the supplier, specify the type of car you have and its body style. If the car has been altered in the firewall or dash area, say so. If he isn't familiar with the space available, get any descriptive literature he may have. Measurements would make it simple. If you can tell him the dimensions behind the dash from your drawings, chances are he can recommend a unit that will fit and cool the car. Plan ahead, it takes a lot of the work out of the project. Rewiring a car or moving a radio can be tough and might make you settle for less than you want.

Selecting a unit that is capable of cooling your car can be tricky. Capacity ratings are a lot like EPA gas mileage ratings, except they can be even more deceptive. You must be sure the manufacturer is actually rating the output of the unit, not just giving a theoretical coil output. Ask local parts dealers or builders. They are pros and usually have experience with components.

A few of the factors that contribute to evaporator performance are air delivery, coil capacity, blower assembly, and case design. The under-dash unit is effective because of its direct air delivery from the coil. When a unit is concealed behind the dash, the air has to find its way out. The simplest and most common way is to run flexible duct hose to dash-mounted vents. Unfortunately, flexible ducting limits the air delivery of any unit because the convoluted wall in the flexhose surface creates turbulence within the hose. Molded ducts are always more effective and produce much quieter air delivery. The difference shows up when you move back from the vent. Flexhose displays a rapid drop-off of air flow, whereas smooth molded duct will produce greater velocity for an extended distance from the vents.

Coil capacity is determined by a number of factors, including coil mass, fins per inch, circuit design, the number of circuit passes. Most in-dash evaporator coils are between 190 and 240 cubic inches and contain between 20 and 60 circuits. Rated capacity of an automotive evaporator coil is based on IMACA Standard 20 and is much more complicated than any one or two factors.

The last major component affecting evaporator capacity is the blower motor and wheel. There is no way for you to check the capacity of the blower assembly, but a few generalities can be stated.

1. The larger, single-blower-wheel assemblies are by far the most advanced today.

2. The larger wheels operate with lower blade tip speeds, making them quieter.

3. The old-style, double-shaft motor and blowers should be avoided. They were not designed for in-dash air conditioners and are very ineffective in conditions when air is being ducted to vents.

OUTSIDE FACTORS

The evaporator removes heat from the interior of the car, but it's just as important to keep additional heat out. A number of things can be done to keep heat from getting inside the car in the first place. Seal all holes in the car. Make sure the floorboards and covers are screwed down tight. Seal all cracks in the car with silicone sealant. Insulate the car thoroughly. Finally, tint the windows (check your local laws regarding the darkness of tint allowed).

When outside air comes into the car, it brings moisture with it, which takes a great amount of energy to remove through the air conditioner. Not only does it make it harder to cool the car, the moisture accumulates on anything cold, creating water problems.

Two other groups of accessories deserve mention. They are pressure and safety switches. A high-head-pressure safety switch cuts the compressor off when the internal pressure is excessive. When the pressure drops to an acceptable level, the switch engages the compressor. The high-pressure fan switch turns an electric fan on when the head pressure rises, and turns it off when the fan removes enough heat to lower the pressure to the minimum pressure level. Finally, the low-pressure safety switch is a safety switch that protects the compressor if R-12 loss occurs.

Cool you should be if you have gotten this far. We have covered all the major components of an air conditioner system, and you should be able to select and install each in an intelligent way.

The facts speak for themselves! The finest street rods built to date prefer Vintage Air. Check for yourself! Why? Because only Vintage Air can satisfy the most demanding customers and situations.

Quality Air Conditioning Parts

Vintage Air is the only manufacturer in the Street Rod Industry that actually manufactures their own complete air conditioning systems. This means Vintage Air air conditioners are designed and engineered in-house to work together as a system. The complete product line forms the famous super cooler and pro-line series - to our econo-line, all perform and last beyond the capabilities of other street rod air conditioning systems.

ALL IN ONE HEAT & COOL For those of you who want maximum capacity with maximum compactness and versatility, choose our All In One Heat & Cooling System.

Factory Trained Dealers

Vintage Air dealers are factory trained. Each year our dealers visit the Vintage Air plant so they learn our product and how to service you - our customer.

Quality Service

Vintage Air attends the major street rod events every season. This way we can meet our customers face to face so that they can see for themselves the things that make Vintage Air products work better and last longer. These shows also give us the opportunity to talk to our old customers to make sure they are completely satisfied with their Vintage Air product. These are only a few of the reasons why, when you know the facts, you will buy the product that sets the standard in street rod air conditioning products - Vintage Air!

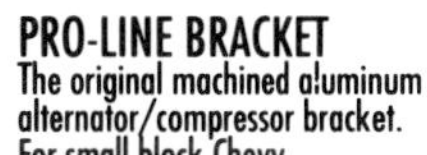

PRO-LINE BRACKET
The original machined aluminum alternator/compressor bracket.
For small block Chevy.

NEW DESIGN THIN-LINE FANS.
Increase airflow when driving slow.
Available with polished motor.

CONTACT ANY OF OUR FACTORY TRAINED DEALERS TO ORDER OR FOR MORE INFORMATION.

ALABAMA
Neil's Wheels
Muscle Shoals, AL
(205) 386-7777

CANADA
The Old Car Center
Langle B.C.
(604) 888-4055

CALIFORNIA
Angie's Auto Center
San Bernardino, CA
(714) 883-8677

Antique Alley
San Diego, CA
(619) 295-4837

Brizio's Street Rods
San Francisco, CA
(415) 952-7637

Buffalo Motor Car
Paramount, CA
(213) 630-6838

California Street Rods
Huntington Beach, CA
(714) 847-4404

Car Custom
Azusa, CA
(818) 969-3596

C.A.R.S.
Santa Anna, CA
(714) 850-9224

Francis (Norm) Street Rods
El Cajon, CA
(619) 588-0349

Frantic Freds Hot Rods
Sun Valley, CA
(818) 504-9307

Hot Rods by Boyd
Stanton, CA
(714) 952-0700

Madsen Auto Air
Concord, CA
(415) 798-1131

Magoo's Auto Shop
Canoga Park, CA
(818) 340-8640

Snow White LTD
Fresno, CA
(209) 255-0527

Mr. Street Rod
Bakersfield, CA
(805) 589-1800

Unique Auto
Montclair, CA
(714) 625-7859

SAC
Anaheim, CA
(714) 680-3373

Frank Schonig
Los Gatos, CA
(408) 353-1407

South Bay Street Rods
Torrance, CA
(213) 325-6667

COLORADO
A-1 Street Rods
Colorado Springs, CO
(719) 577-4588

The 40 Fort Street Rods
Arvada, CO
(303) 420-9800

Masterpiece
LaFayette, CO
(303) 666-7637

CONNECTICUT
Total Performance
Wallingford, CT
(800) 243-6740
(203) 265-5667

Barry's Garage
Preston, CT
(203) 889-7684

FLORIDA
J & B Custom
Stuart, FL
(407) 220-2875

Mark's Air
Tampa, FL
(813) 988-1671
(800) 521-9854

Street Rods Unlimited
Pensacola, FL
(904) 479-7637
(800) 828-7637

Willie's Street Rods
Hudson, FL
(813) 862-4659

Dana's Hot Rod Shop
Ft Pierce, FL
(305) 466-1932

GEORGIA
Joe Smith Automotive
Atlanta, GA
(404) 634-5157

Yank's Rod & Performance
Cumming, GA
(404) 781-9673

Frank's Auto Clinic
Liburn, GA
(404) 564-1710

ILLINOIS
Frame-Up Wheel Works
Gurnee, IL
(312) 336-3131

Midwest-Street Rods
Boling Brook, IL
(708) 739-1940

Roadster Shop
Elgin, IL
(708) 742-1932

Rockvalley Antique
Stillman Valley, IL
(800) 344-1934

INDIANA
Russell Custom Rods
Dayton, IN 47941
(317) 447-4700

IOWA
Classic Carriage Works
Des Moines, IA
(515) 262-9486

Yogi's
Wyoming, IA
(319) 488-3111

KENTUCKY
Plumformance Custom
Louisville, KY
(502) 969-1370

MARYLAND
Anderson Industries
Elkridge, MD
(301) 796-4382

MASSACHUSETTS
Belmont Rod & Custom
Dedham, MA
(617) 326-9599

MICHIGAN
Jerrie's Custom
Metamora, MI
(313) 678-2796

Vintage Rodder
Burton, MI
(313) 742-8587

MINNESOTA
Metal Fab
Blaine, MN
(612) 786-8910

Twin City Rod & Custom
St. Paul, MN
(612) 774-6470

MISSOURI
K.C. Street Rods
Kansas City, MO
(816) 781-0777

Pete & Jake's
Grandview, MO
(816) 761-8000
(800) 344-7240

The Wherehouse
St. Louis, MO
(314) 351-7533

MONTANA
Dagel's Street Rods
Helena, MT
(800) 325-3149

NEBRASKA
Cornhusker Rod & Custom
Alexandria, NB
(402) 749-7565

Gibbons Fiberglass
Gibbon, NE
(308) 468-6178

NEVADA
Hot Rods of Las Vegas
Las Vegas, NV
(702) 736-4005

NEW JERSEY
Custom Auto Radiator
Jackson, NJ
(201) 364-7474

Hot Rod Shop
Elmwood Park, NJ
(201) 796-8807

NEW YORK
Super Rod Shop
W. Babylon, NY
(516) 643-5968

Yesterday & Today Auto
Mineola, NY
(516) 746-6880

Heritage Auto
Pawling, NY
(914) 855-5388

NORTH CAROLINA
Johnny's Rod Shop
Mooresville, NC
(704) 663-1568

OHIO
Butch's Rod Shop
Dayton, OH
(513) 298-2665

Lobeck's Hot Rod Parts
Cleveland, OH
(216) 439-8143

Springfield Street Rods
Springfield, OH
(513) 323-1932

Ohio Valley Street Parts
Cincinnati, OH
(513) 574-9217

OKLAHOMA
Parr Automotive
Oklahoma City, OK
(405) 942-8677
(800) 654-6779

OREGON
Kenterprises
Salem, OR
(503) 364-0967

PENNSYLVANIA
J. P. L.
Beaver, PA
(412) 775-1932

Outlaw Performance
Avonmore, PA
(412) 697-4876

Posie's Horseless Carriage
Hummelstown, PA
(717) 566-3340

California Street Rods East
New Castle, PA
(412) 658-5510

R-T Street Rods
Coopersburg, PA
(215) 282-1726

Queens Auto
Macungie, PA
(215) 966-4310

TENNESSEE
Eddie's Rod Shop
Kingsport, TN
(615) 323-2081

TEXAS
Alamo Hot Rod Parts
San Antonio, TX
512-732-1940

Sachse Rod Shop
Sachse, TX
(214) 495-1557

Vintage Friends
Humble, TX
(713) 540-2828
(800) 999-2890

Yearwood Speed & Custom
El Paso, TX
(915) 772-7408
(800) 444-3331

WASHINGTON
Greer Enterprises
Richland, WA
(509) 627-3411

R.B. Obsolete
Snohomish, WA
(206) 568-5669

Vintique Reproductions
Tacoma, WA
(206) 537-8261

10305 I.H. 35 N., SAN ANTONIO, TX 78233

1-800-423-4172
(512) 654-7171

POWER WINDOWS

Power windows are very convenient accessories. Late model electric window lifts can be modified to work with early windows, but if a single lift arm is used, it may be necessary to keep experimenting with the motor location until the glass moves freely without binding. Some aftermarket window lifts are available that work in a pure vertical fashion. Others have the more traditional arm.

Really cool is the idea we picked up at Gene Winfield's shop (but an idea that's being used all across the country, on both rods and customs). The entire door inner structure, including power lift mechanism from a late model car door, is cut out and grafted into the early car door.

In the accompanying photos, a Mercury door has been relieved of its inner structure in preparation for receiving a late model unit. Of course, it is necessary to do a little tape measure work to figure out which late model door will mate with the early door. But generally, GM car doors of the 1970s and '80s can be made to fit, with a bit of trimming. If the donor car door was equipped with a guide for curved glass, it must be replaced with a straight guide (homemade) that will position the early car flat glass.

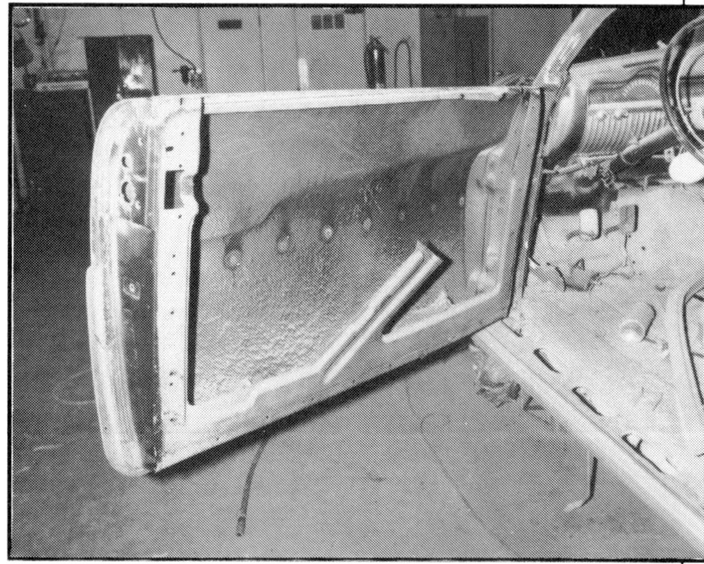

This Mercury door has been cut open and the inner structure removed. Just enough of the lip was left to maintain rigidity while working with the new insert.

The inner panel from a late 1970s-'80s GM car door was cut and trimmed until it fit the Mercury door. Select a donor car that has doors which are very nearly the same size and shape as the Mercury or shoebox Ford doors. The late model door will probably have curved guides for the curved windows. These must be replaced with homemade straight guides for locating the straight glass of the early cars.

Bob Beverly solved the power window problem in another way. Wanting to install electic power units in his Merc, he opted for hardware out of a '76 Cadillac. Some modification was necessary to make the linkage function, and braces needed to be fabricated, but this is the kind of job that is well within the capability of many rod and custom enthusiasts.

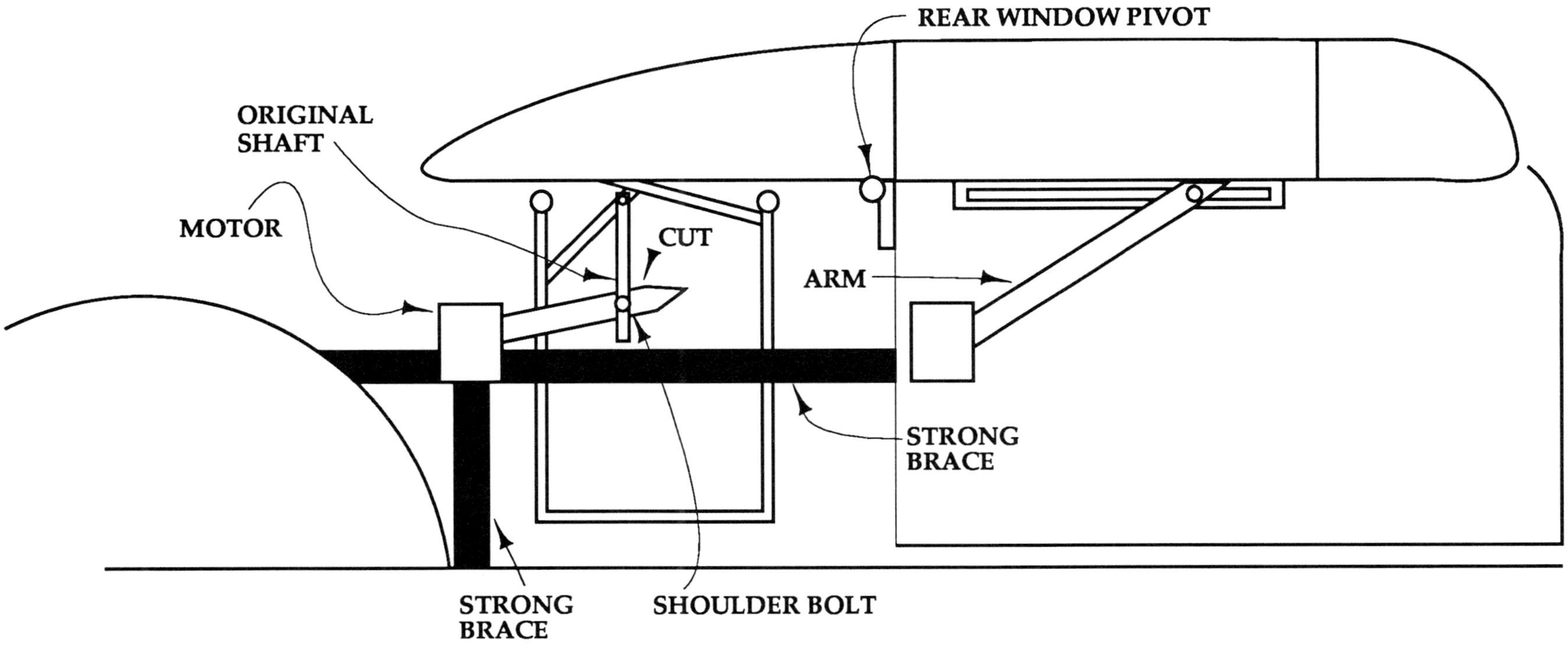

REAR WINDOW PIVOT

ORIGINAL SHAFT

MOTOR

CUT

ARM

STRONG BRACE

STRONG BRACE

SHOULDER BOLT

Front - For the front windows, Bob Beverly used a '76 Cadillac motor and arm. The Cadillac track assembly was mounted on the bottom of the Mercury window. The Cadillac motor was mounted by drilling holes in the inner door panel.

Rear - The spring and hydraulic cylinder were removed. The arm of another '76 Cad power unit was shortened and attached to the shaft of the original cylinder, which hooks to the window with the original linkage. Braces to mount the motor assembly must be fabricated.

by Rich Johnson

CUSTOM TAILLIGHTS

Jim Ballard is every ounce our kind of guy. Just feast your eyes on what he's been up to with his '53 Mercury. The project isn't finished yet, but that doesn't really matter. What does matter is the kind of work Jim has been doing to this car.

The first time the car came out of the garage in 2-1/2 years, was to take in the '90 version of the Goodguys meet at Puyallup, Washington. Jim was feeling kinda bad because his ride was unfinished, but any embarrassment swiftly disappeared when people spotted the Merc's 1990 Cadillac Fleetwood taillights. Lots of folks said these were the neatest taillights they had ever seen in a '52-'54 Merc. They just stood there and stared at them!

Of course, this is just one of an infinite number of ways to customize the taillight area of a car. But this general approach can be employed for any number of different custom taillight installations. By following along through the photos and captions, it's easy to see what is involved in radical taillight surgery.

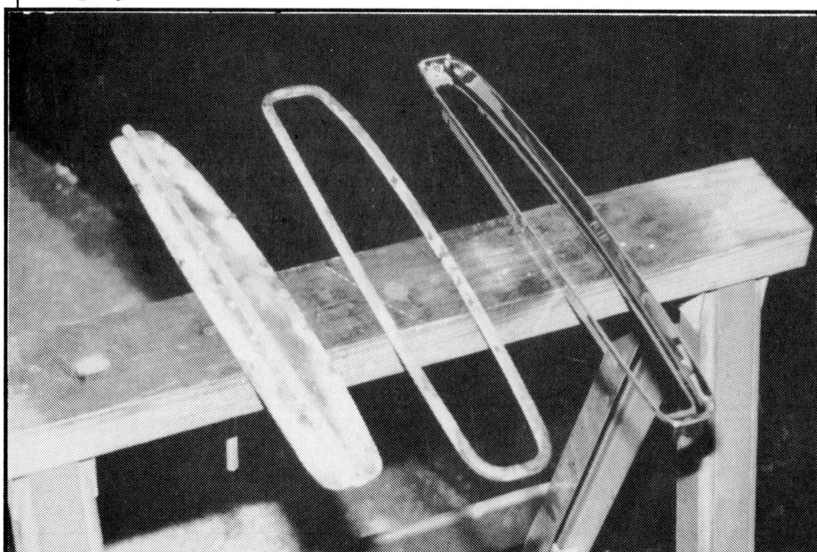

The hardest part was forming the 5/16" rod frame to perfectly fit the compound curves of the chrome taillight bezel. The problem was solved by making an exact size and curvature template of the bezel out of 16-gauge sheetmetal, and reinforcing it with some 1/4" rod. With the template mounted in a vise, Jim formed the 5/16" rod frame by tack welding the rod to the template with a MIG and then heating the rod with a torch to shape it around the template while tack welding every couple of inches. When finished, the rod frame was cut loose with a die grinder, then ground smooth. Short tabs were welded on for screw clips to accept the stock Cadillac screws.

Left-When locating the taillights on the back of the car, it was discovered that when they were positioned exactly vertical, they appeared to lean out at the top, due to the curvature of the body sides. It was necessary to lean each taillight in at the top about 1/8" so they would appear to be vertical.

Above-Jim made a couple of sheetmetal box structures that extend into the Merc fenders a few inches through the stock taillight fender openings. These boxes were tack welded to the fenders and completely sealed around all of their edges. This effectively isolates the Cad taillights from the trunk compartment, to eliminate any possibility of water leaks into the trunk from the taillights.

Cadillac wanted $110 for the rear wire harness with the 6 light sockets. These sockets are a different style from other late model GM cars. After searching a very large GM wrecking yard, Jim finally found a perfect fit in an older Olds Toronado for $10. Before doing any work or welding on the bumper, it should be removed from the car, and the brackets disassembled to relieve any built-in stress. Jim's bumper didn't want to go back on in its original position, and it was a lot of work getting it to fit right again.

The 16-gauge filler pieces for the bumper cutouts were farmed out to a sheetmetal shop for $10. It is important that these fillers exactly match the curve of the 5/16" rod framework. Jim allowed a gap of about 3/16" between the taillight frame and the bumper filler, but 1/4" would have been better. This gap can be filled with some type of weatherstrip rubber for a more finished appearance.

Talk about a slick looking '53 Merc! The '90 Cad taillights look like they were made for this car. Total cost of this taillight project, including new taillights from a Cadillac dealer, sheetmetal, 5/16" and 1/4" rod, bulbs and sockets, stripping the chrome off the bumper, bondo, primer, etc. was less than $225. Jim then sold his stock taillights at a swap meet for $100, so the final cost was about $125.

by Rich Johnson

FLUSH DOOR HANDLES

Jim Ballard performed a slick trick on the doors of his '53 Mercury. Choosing a set of junkyard '87 Honda Prelude door handles, Jim worked over the mechanism slightly and the result was a smooth operating, clean looking installation. At the first show Jim attended, the shoebox Ford crowd seemed especially interested in this flush handle installation. Several mentioned that they wanted to get rid of their solenoid operation, and these handles are readily available for a reasonable price.

Honda Prelude door handles are the only readily available units Jim could find that were totally flush mounted, which makes them desirable for customizers. Wrecking yards know this, and price them accordingly. There are cheaper handles, but they don't fit absolutely flush. (They would be easier to install, though. Just cut a hole in the door and bolt them in.) Total cost for these two handles was $115.

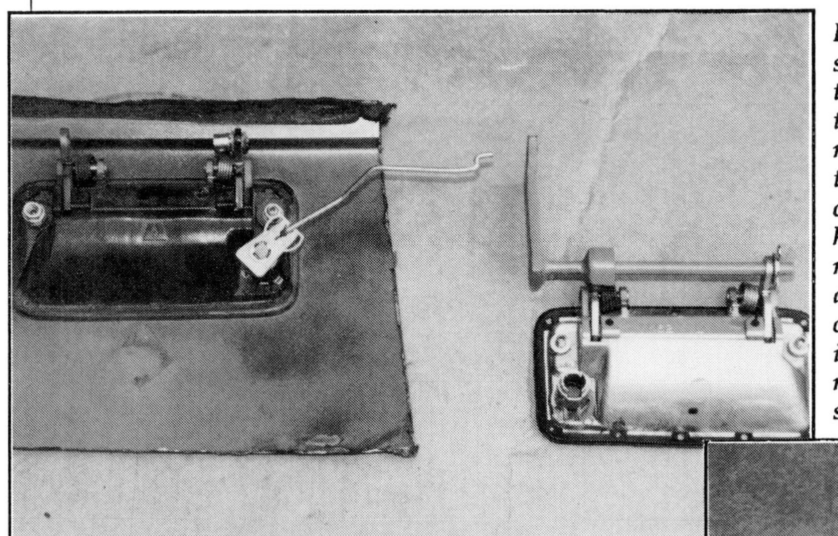

Left-The most important point to watch for in selecting a handle is to match the curvature of the donor door to the door you intend to install the handles in. Then make sure the actuating mechanisms are compatible with each other. In the case of the Merc/Honda combination, it couldn't have been more perfect if the factory had designed it. Connecting the key lock mechanism to the Merc is quite simple. There is a plastic part on the back of the Honda key cylinder. This is removed, and a duplicate piece is made up of metal (for strength). Then a metal rod or pin is welded on, which will engage the stock Merc door lock mechanism in the door.

Right-The mechanism was extremely simple to make. Some 5/16" rod, a little flat stock, a washer and cotter pin was all it took. Installing these units in the Merc was simple to the point that they work even better than the original handles.

Left-After the Honda handle was removed from the donor door, the Merc door was marked and a large enough hole was cut to accommodate the handle and an inch of surrounding sheet-metal. The better the hole matches the piece to be installed, the less difficult the finish bodywork will be later on.

Right-Here's the finished product, and it looks as clean as any factory could do. The Honda Prelude flush-mount door handle is an ideal late model custom upgrade for this vintage Merc.

by Rich Johnson

PRO STREET

Vehicle: 1949 Ford Pro Street
Owner: Don Crismore — Albuquerque, New Mexico

Does this look like one bad shoebox, or what? With a front ground clearance of only 2-1/2" and a built rat inside the engine compartment, Don Crismore's box definitely turns some heads. And when it's all finished, it's likely to turn some quick times at the stoplight drags.

Before the floor and firewall were completed, the engine and TH400 transmission were set into place to check for fit. This was necessary to get a general idea about the shape the floor and firewall would need to take. At this early stage, the doors were still opening in stock fashion.

Top to bottom, end to end, this is one bad box! Not exactly the kind of Ford a little old lady would want to drive, unless of course she was wearing a Nomex suit and lived for the aroma of fried tires. But then, Don Crismore is no little old lady, either.

Don decided that after four decades of mundane life in the New Mexico desert, this shoebox deserved to be treated to a healthy shot of steroids. Fortunately, life in the desert was kind to the sheetmetal, so there was virtually no rust to deal with, although a bit of damage had occurred in the cowl and adjoining floorpan area. This presented little problem, because for what Don had in mind the entire floor and firewall had to come out anyway.

Building a pro streeter is a whole 'nother ballgame from doing a restoration or simple hot rod. Massive chassis work has to be done to accommodate things like monster motors and gargantuan rear wheels/tires. Gary Burns of SWEB Enterprises in Albuquerque did all the chassis fabrication.

At the rear, the frame rails were cut off at a point just where the rails begin to rise over the rearend. A TCI crossmember with a driveshaft loop was installed, then coilover shocks and a 4-bar setup were employed for the rear suspension system that is attached to the narrowed (39" overall width flange to flange) Ford 9" rearend. The TCI crossmember was welded to the rear portion of the severed frame, and the extracted

Note how much firewall work was necessary to prepare this shoebox to receive the monster motor that was to come. Essentially the entire firewall and floorpan had to be removed and custom made to fit. Mustang II front end serves suspension, steering and disc brake duty. By this point, the suicide doors were completed.

With front sheetmetal missing, and a pop can poised by the frame rail for size comparison, it's easy to see that speed bumps are out of the question. The 464 rat motor has been treated to a 6-71 blower, and there's plenty of shiny stuff to dazzle the eye.

A TCI rear crossmember was welded into place and a driveline loop built in. Rear suspension consists of a set of coilover shocks and 4-bar setup. Note how the stock Ford frame was cut just where the kickup began, then the rails were narrowed to make room for all that rubber.

pieces were then narrowed, allowing clearance for a set of 31x28.5x15 Mickey Thompson tires and Cragar Drag Star Super Lite wheels. At the very rear of the frame rails, the body was reattached and the bumper mounting brackets were put back in their original positions.

The floor and firewall were completely removed and replaced with 18-gauge sheetmetal. Don says that he thinks they could have used 20-gauge with as much success, and certainly less muscle to bend the tubs. Having had no previous experience with a MIG welder, it took him awhile to get the job done, but it looks good and works just fine.

Up front, the frame was severed at a point where the firewall meets the frame. The main frame (the part under the floorpan) was then set on 4x4's to get the desired ground clearance, and the front end was fabricated to this height. It's super low! Rectangular steel tubing (2x4) was used for fabrication of the new rails, and a Mustang II front end was installed. Anchor brackets for the bumpers and radiator were then reattached in the proper locations. A set of 2" dropped spindles are employed to get the front end down to a total ground clearance of only 2-1/2" with the front bumper installed. Front wheels are Cragar Drag Star, bolted to Outlaw 5-lug rotors.

For power, a 464 cid rat engine was stuffed between the fenders and then topped with a 6-71 supercharger. Getting through the gears is the job of a TH400 automatic transmission, and the axle ratio is a quick 4.11:1.

Modifications to the body were delegated to John Harvey of Lil' John's Kustom Studio in Los Lunas, New Mexico. To smooth out the exterior, several techniques were used. The windshield wipers were removed and the holes filled. Door handles were discarded, and the trunk hinges were removed and replaced with inside hinges from a 1951 Ford. Rear fender metal welting was eliminated and the fenders smoothed. Rear window opening was shrunk to within 1/8" of the glass measurement, and S-10 molding used. The same will be done with the V-butted windshield. Side quarter windows are to be flush mounted on the outside of the metal lip, and the door glass is a one-piece unit, eliminating the vent windows. Electric window lifts replace the hand-crank regulators, and this system will serve as the means of getting into the locked car, rather than using electric door latch solenoids. As if all that was not enough, Don has designed the car with suicide doors, utilizing '57 Chevy hinges and VW type III latches.

If you're ever cruising the southwest, look for Don. He's a member of the Duke City Rodders, and he drives one bad shoebox.

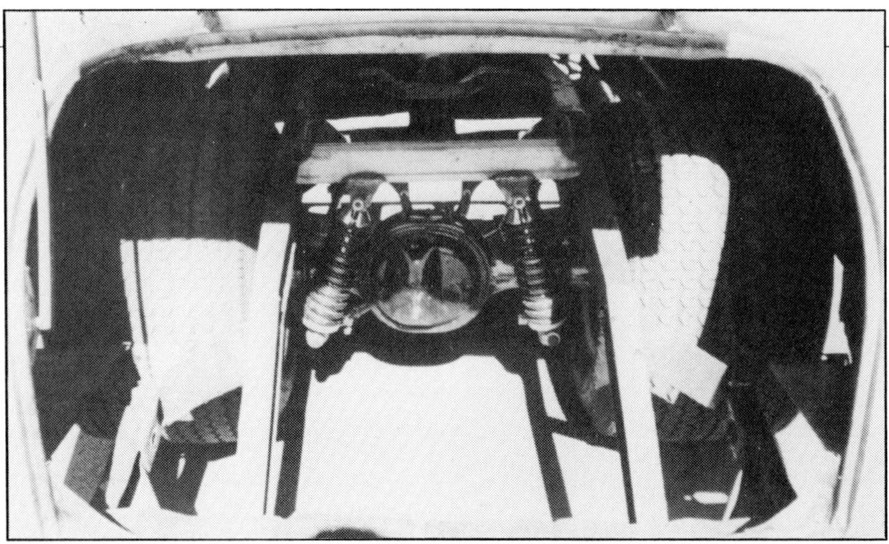

A narrowed Ford 9" rearend found a permanent home between a pair of massive Mickey Thompson tires and Cragar wheels. Overall width of the rearend, measured from flange to flange, is only 39".

Note the use of '51 Ford inside trunk hinges, chosen to replace the stock outside hinges in order to give the car a smoother look.

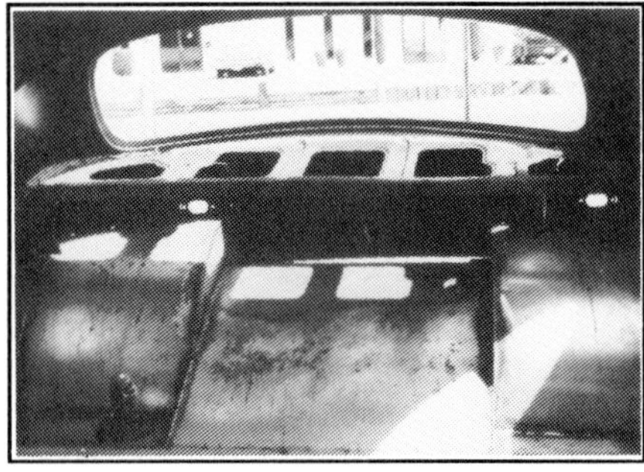

The new floorpan was built of 18-gauge sheetmetal, although Don says they could have probably gotten along just fine with lighter weight 20-gauge stock.

Door latches are VW type III units, which fit very cleanly and do the job. With creative thought, some skill, and a ready supply of salvage parts, anything can be done with a hot rod.

Don wanted suicide doors, and came up with a plan for building them. For hinges, he selected a set from a '57 Chevy, and they fit nicely. Note the additional tubular bracing that was installed to support the hinges.

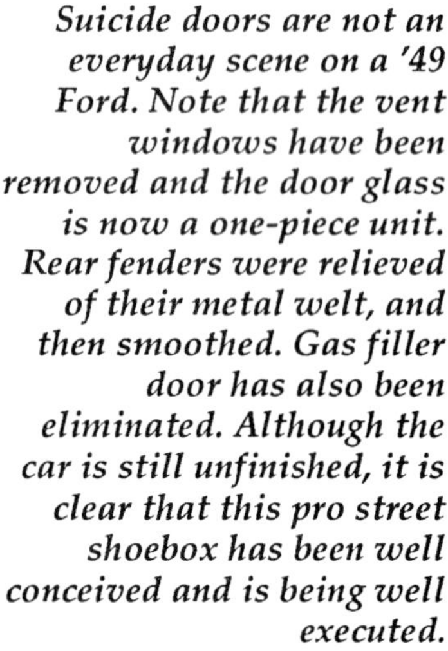

Suicide doors are not an everyday scene on a '49 Ford. Note that the vent windows have been removed and the door glass is now a one-piece unit. Rear fenders were relieved of their metal welt, and then smoothed. Gas filler door has also been eliminated. Although the car is still unfinished, it is clear that this pro street shoebox has been well conceived and is being well executed.

by Tex Smith

HIGH SCHOOL HAULER

I suppose the neatest thing about a shoebox, be it Ford or Merc or any other 1950's era car, is that the styling is both old and modern. Unlike the more traditional hot rods, those designs of the 1930's that are in every respect pure antique, the shoebox is a perfect transition piece. It has a more modern suspension and chassis, the engines are rather contemporary, and the body styling is somewhere between "then" and "now." Size-wise, a typical shoebox ranks right in there with many of the mid-size cars of today.

Which is why this particular Ford is such a sweetheart. Yes, we are talking about two different cars here, but I'll start with just one.

I used to hang out at a great North Hollywood shop that specialized in engine swaps (Herbert & Meek Automotive). And from that shop, I did hundreds of swap articles for magazines such as *Hot Rod, Car Craft, Rod & Custom*, etc. Don Herbert had been in hot rodding for years, and his son was coming into the business, working weekends and after school. Andy needed a car for school, so when a nice 1949 two-door Ford showed up, he grabbed it. This was a time when late '50s Chevys were everywhere, as were GTO's and hot 4-4-2's. Van Nuys Boulevard was a favorite cruising spot for southern Californians, and you would think that Andy would go for one of the muscle era cars. It was a surprise that he chose the shoebox, even if his shop did cater to every conceivable engine swap possible.

The interesting result was more of a surprise. At high school, and on Van Nuys Boulevard, Andy's small block Chevy powered boxcar proved to be extremely popular. The car was old enough (about 15 years old at the time) to get lots of stares from his peer group. And with the updated engine, it was strong enough to put some factory pretenders on the trailer. Anyway, the car was a champion. Our photos show why.

As so many young men in southern California did, Andy Herbert mixed a strong high school technical education with a hobby and a future career. His choice for transportation was a Chevy V8 powered 1949 Ford two-door sedan. Although it could be called a resto-rod, the car had such pleasing road manners it quickly became popular with all age levels.

American Racing mag-type wheels gave a wider than stock stance to Herbert's ride, which drastically improved handling. Note how the wider tires fill the wheel wells nicely. The front end was not lowered.

Original trim was left in place. The car was painted black, and the effect was highlighted by the custom wheels. Car enthusiasts immediately identified the car as a "sleeper." Regular tourists thought it was a new model.

Small block Chevrolet V8 is a perfect swap for the shoebox Ford. Ignition clears the firewall nicely and stock Ford radiator is more than sufficient to cool the engine. Stock Ford transmission can be used, with adapters.

The stock Ford interior takes well to simple improvement in upholstery material. There is plenty of room for A/C and stereo additions.

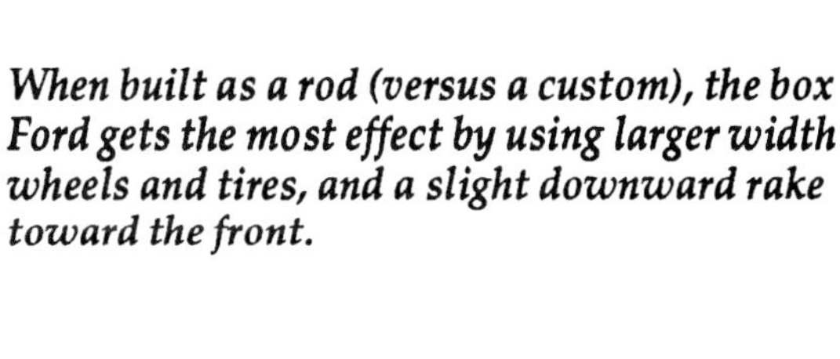

When built as a rod (versus a custom), the box Ford gets the most effect by using larger width wheels and tires, and a slight downward rake toward the front.

by Tex Smith

HUGO THE UGH!

Andy Herbert kneels in front of his 1949 Ford street hauler, while car enthusiast friend discusses building a clone. When these cars were built, during the 1960's, shoebox Fords were seldom (if ever) considered hot rod or custom projects.

Several years after the previous photo was taken, the 1960 two-door was rescued from Herbert's yard. Minor damage had been caused to the grille and trunk lid, but the pleated vinyl interior was still good and the paint was still in excellent shape.

If you haven't done so, go back to the preceding two pages and read about Andy Herbert's really keen 1949 Ford two-door high school hauler. That car has a direct connection with the 1950 two-door we talk about here.

You see, while in high school, Andy worked at his dad's automotive repair shop, a business that was world famous for doing engine swaps. From very early, Andy wanted to be an auto mechanic, so when he got to North Hollywood High, he concentrated on the tech training end of things. And he belonged to the local car club. His '49 became one of the most popular cars in the vast school parking lot, where a friend saw the car and decided on one "just like it." Well, not exactly the same, because the friend found a 1950 model. Close enough, though, that when the cars were finished (both painted black), they were always being confused with each other. Andy and his buddy thought that was neat.

Several years passed, and one day Andy (in his twenties) asked me if I would like to have a friend's '50 for my son Scott to drive to college. Andy's buddy had set the car aside a few years earlier. The small block Chevy V8 had been removed, but otherwise the car was in good condition. To sweeten the deal, Andy had a nice 327 small block that he had built for another purpose. Hopped up, but just right for the street.

Momma didn't raise no dummies at our house, so

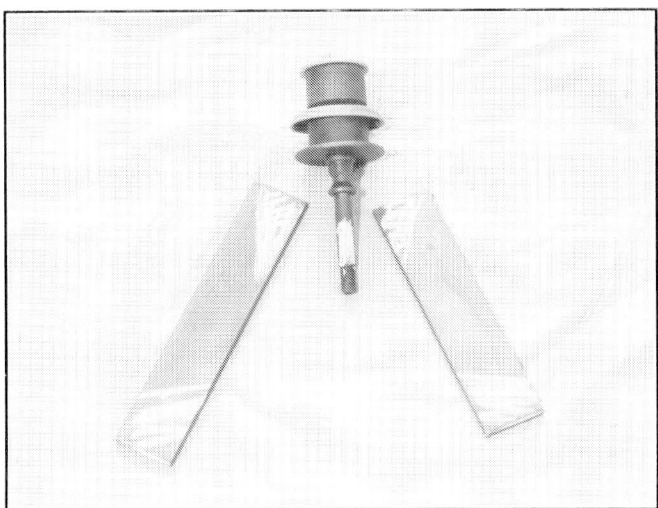

Working with a Chevy block that can use the engine mount bracket off the front of the block, these are the pieces that are used to create new mounts. There are lots of ways to make mounts, but this seems to be the most simple.

It is still possible to find shoebox Fords and Mercs in great driveable condition, especially in the dry climate states. But these cars have taken on new popularity in recent years, so today's prices are working upward. Still, the Ford does not command as much as a two-door Merc.

we scooted right down to Andy's and got the car. While the body/chassis was being cleaned up (very little needed), we had a special PowerGlide transmission created by Al out of Van Nuys, California. Originally, the '50 had used a standard Ford transmission behind the Chevy engine, but we wanted something a bit smoother. With the trans bolted to the engine, we dropped the unit into the car and noted that making the transmission mount would be very easy. Took about an hour, and the engine was in place.

But, we're ahead of the story. The reason the engine fit, was that Andy and his buddy had installed motor mounts years before. They had also modified the exhaust manifold to clear the Ford steering box. We show you these items as an explanation of how really simple a Chevy/Ford boxcar swap can be.

While Scott was working on the car, his sisters nicknamed it Hugo The Ugh! They though it was the ugliest car they had ever seen. Being Valley Girls, they were certain that no one ever drove anything but a BMW. Scott ignored them, and ended up with a really nice running and handling road machine. It got over 20 mpg, it would most definitely scoot across the southwest desert highways at well over 100 mph, and as Andy had learned years before, the car was very popular with the people who counted. The hot rod set.

A few years later, Scott decided he liked my Chevy El Camino, so we traded. Hugo's neat engine was switched to the Chevy, and Hugo took up residence in the back yard for about a year, until Wayne Atkinson up in Provo, Utah called and asked about the car. Yep, I said, it was available. But only to someone who would appreciate what it was. Mike Petersen in Salt Lake City was the someone Wayne had in mind. Mike drove to our southern California home, towed the engineless car home, and I forgot about it until recently. Now, all these years later, Mike still has the car.

Hugo The Ugh! lives!!! Not much the worse for wear, and someday he'll probably be back on the streets. Proving for sure that you can't keep a good shoebox car down.

Pieces of steel plate are welded to the rubber biscuit insulator, then to the front of the Ford crossmember.

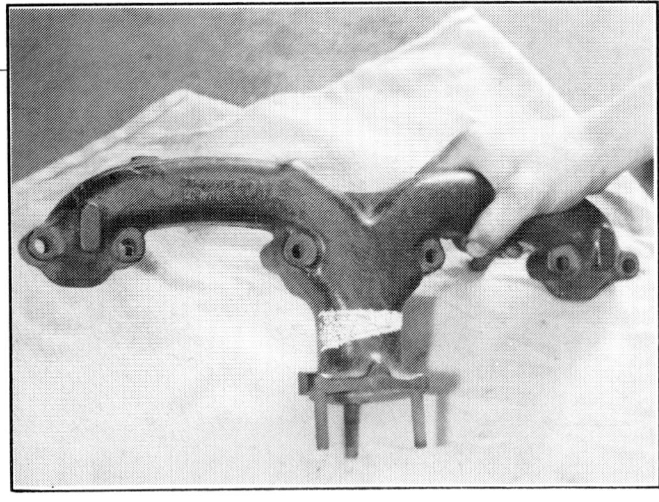

The ram's horn exhaust manifold for the driver's side must have the outlet angled to clear the Ford steering box. Cut out the shaded area, weld, and the manifold fits.

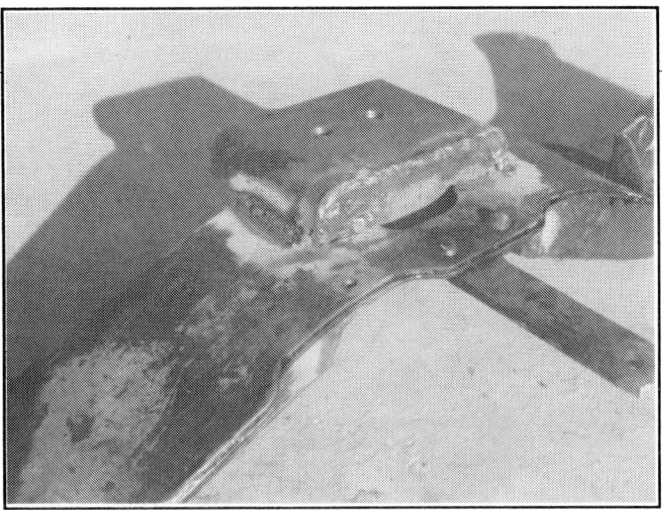

If a PowerGlide automatic transmission (or later model TurboHydro) is used, a mount must be welded to the stock Ford crossmember.

When the Chevy transmission is used, a new driveshaft must be made up to mate with the Ford rearend. Yes, the shoebox Ford/Merc rearend is strong enough for most modern ohv V8 engines.

When the small block Chevy is mounted, the alternator will often interfere with the passenger side splash panel. The solution is to make a small box in the panel for alternator clearance.

There will be more than enough clearance between radiator and fan, so a fan extension is recommended. In addition, a radiator fan shroud can be adapted, should the car run hot. This was never a problem with this particular car, however, even on extremely hot desert days.

by Calvin Mauldin

SMOOTH '49

Vehicle: 1949 Ford Club Coupe
Owner: Michael Wheeler -- Dallas, Texas

Definitely keeping a low profile, Michael Wheeler's '49 Ford club coupe features a 3-1/2" top chop. Relaxing the hydraulics puts her down on the pavement.

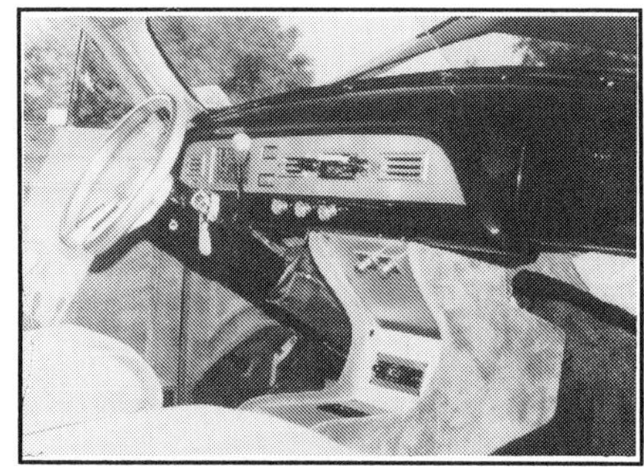

Modified '51 Ford dashboard has been filled with all the latest goodies. The steering wheel is a handcrafted one-of-a-kind. Console holds the controls for the hydraulics and power windows, plus the radar detector.

SMOOTH (smooth) adj. 9. Without lumps; having the elements perfectly blended; a smooth mayonnaise. That is Funk and Wagnall's definition of smooth. In the world of custom kemps, Michael Wheeler's Texas-based '49 Ford coupe is the ultimate rolling definition of the word smooth.

Modifications to this Fo/Mo/Co product of the shoebox era run the gamut of the traditional customizer's handbook. Starting with a 3-1/2" top chop to get things looking right, the original builder (Bob Sipes of Grandview, Missouri) frenched the headlights and tunneled the taillights, did a nose and deck job, shaved all the chrome, and molded in a '54 Pontiac grille. That's a pretty good day's work for anybody, but Bob had a few more tricks up his sleeves. Like punching 120 louvers into the hood, for good measure. Up to this point, what Sipes had was an old-style '50s custom, but then he went one step beyond.

Designed with long distance cruising in mind, the interior details read like the option list for a new luxury car. Appointments include tinted power windows, power steering, power brakes, 6-way power seats (modified Corvette units), AM/FM cassette sound system, built-in console with radar detector, air conditioning, and one-of-a-kind handbuilt steering wheel. On the outside, Ditzler Brocade Rose Pearl was applied to the ripple-free body by Roger Ward of Ottawa, Kansas.

Mechanicals were updated with a Nova subframe, a very warm 350 Chevy engine, TH400 automatic transmission, 4.11:1 rear, and a complete set of hydraulics front and rear.

Under the punched hood, the engine room is made up of a dazzling display of polished stainless steel. The stainless, and the super-detailed small block V8 are truly works of automotive art.

The overall theme of the forty-niner works well, mixing the old with the new, no lumps, perfectly blended. Smooth. Maybe even smoooth.

The much-massaged hood is punched full of louvers -- 120 of 'em to be exact. Windshield was butted to give a clean one-piece look.

Front end treatment gets an A+ for styling and execution. 1954 Pontiac grille and smoothed bumper are perfect touches for a shoebox Ford. Headlights were frenched, using '52-'53 headlight rings.

Customizing the rear of a '49-'51 Ford can be quite a challenge. Not radical, but very eye-pleasing are the tunneled stock taillights and the frenched license plate.

Low, deceptively simple, and smooth. Makes us wonder why Dearborn designers put all that chrome stuff on in the first place. The flawless body was treated to Ditzler Brocade Rose Pearl by Roger Ward.

It's the little tricks, like the gas filler door, that make you appreciate all the hours that went into this custom. Shoebox Ford experts will remember that the gas cap was exposed on the stock '49.

Better be wearing your Foster Grants when you pop the bonnet on this shoebox, or you might go shine-blind. Craftsmanship and detail of the stainless engine room and small block Chevy V8 are astounding. Consider too, this is not a display-only showpiece. Wheeler drives it everywhere!

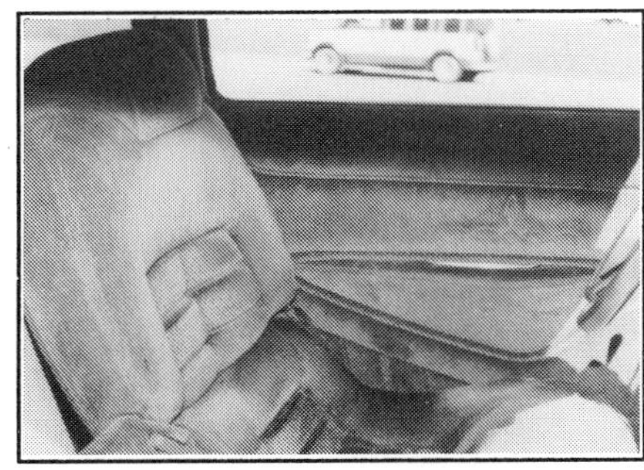

Modified Corvette 6-way power seats make those long highway miles easy on the backside.

Fully upholstered trunk conceals all the hydraulic hardware. Trunk support looks like it could be utilized as a multi-purpose piece.

By Jim Genty

THE KID

Vehicle: 1949 Ford Business Coupe
Owner Scott Pfitzinger—Ballwin, Missouri

Smooth flowing lines of the old Ford represent old and new in a 1990s kind of monochromatic statement. Can this be the start of a new youth movement?

Growing up around his dad was pretty much the same for Scott Pfitzinger as it was for any other kid. He did all the normal things a kid does, Little League, soccer, welding, painting, and engine rebuilding. Well, maybe not all the same things.

The love for custom cars just naturally rubbed off on Scott as he watched and learned from his dad, Jerry. When Scott reached adulthood, they felt it was time to work together. Jerry showed pride in his son's talents and soon people were saying that Jerry and Son Automotive was the place to go in St. Louis. As time progressed, Jerry found himself saying over and over again, "Bring it down and let the kid look at it."

Since dad already owned one of the hottest looking customs in St. Louis, Scott just naturally followed in his shoes. No mini truck for this kid, he was going to build a custom.

Scott located a 1949 business coupe and jumped in with both feet. He took it to Danny McDowell of McDowell Auto Body in St. Louis, and had him relieve the top of 6" of unwanted height. With that out of the way, he brought it back to the shop and started to apply what he had learned from his dad.

His first priority was to get the car down, but not with hydraulics, he wanted it done right. He bought a set of 3-1/2" dropped spindles from Gene Winfield, and installed them with new coils. He rebuilt the entire front end, added KYB gas shocks, and then dropped the steering arms and drag link. This brought the front end down on the ground, but afforded him a decent ride. But now was the drive shaft rubbed...no problem. Scott fabricated a new tunnel that is 3" higher and runs the entire length of the floor.

Out back he C'd the frame kickup another 3" (a trick he learned from dad) and narrowed the rearend. Since no one in St. Louis was narrowing rearends, Scott set up his own jig in the shop and did everything himself, including cutting and resplining the axles (who says that hot rodding is dead). Since he wanted to be able to keep up with late model Camaros, he added heavy duty springs from a Maverick to keep the P245/60 tires glued to the ground under hard acceleration. Low, but not slow, was his theme.

It was now time to turn his attention to the rest of the body. He shaved the door handles, filled the trunk and removed the outside hinges. Up front he shaved the hood and added a full-length peak, complemented by six rows of louvers, totalling over 120. For the grille area, he molded in a '51 Ford top bar and created a unique custom grille from '51 Dodge ends attached to a '56 Plymouth Savoy center section.

To keep with the smooth trend he had started, Scott located a pair of original '49 Plymouth bumpers and molded them in front and rear for a really striking effect. Out back he cut a half moon radius on each side and tucked the stainless steel exhaust tips up close for clearance on those nasty driveways.

The stock taillights and headlights were frenched in and the body was prepped for paint. Scott felt that he wanted an award winning paint job to go with all the fine body work on the car so he called on the gang at Donovan Auto Body. There they applied Ditzler Porsche Red urethane to everything, including the spot lights and Enkei directional wheels.

After the car was rubbed out, Scott took it back to the shop where he installed a motor he had built from a '72 350. The small block is now 355 cid and is topped off with an HEI ignition, Edelbrock manifold with a reworked Thermo Quad carb. Inside there is a mild cam and head work by the owner. Scott backed this up with a Jerry and Son modified Turbo 350 for those positive shifts.

Ready for street or strip, it should be mentioned that this car is as Scott intended, low but not slow. Scott turned a 13.2 ET in 1,000 ft. at a recent Nostalgia Drag Race in the St. Louis area.

It is evident that "like father, like son" does prevail at the Pfitzinger house. And while dad still does have an outstanding custom, he has to really put the pedal to the metal to keep up with "The Kid."

Engine compartment glistens with new paint, wiring and modern day motivation. Notice the attention to detail.

At the rear we see the '49 Plymouth bumpers that Scott molded into the body. Note the exhaust cut out on each side for the stainless steel exhaust tips.

Early construction shot shows the professional approach McDowell Auto Body took to dropping the top and slanting the door posts forward.

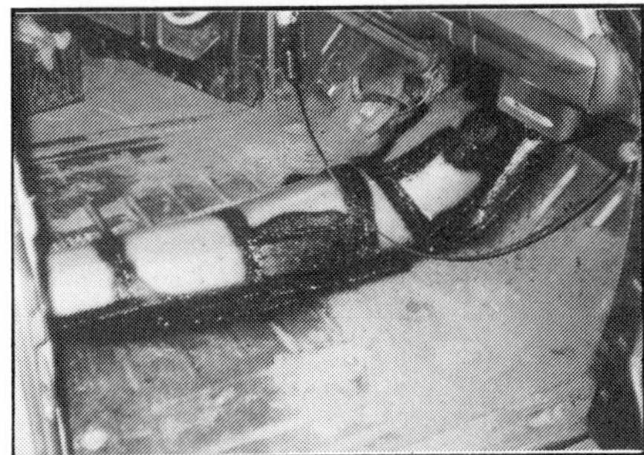

An all new transmission tunnel was built by Scott that runs the entire length of the interior floor and gives the drive shaft an additional 3" of clearance.

This was Scott's first outing at the Nostalgia Drags last year. The little shoebox turned a very respectable 13.3 ET at the 1,000 ft. mark.

by *Jim Genty*

PASSION PIT

Vehicle: 1950 Ford Tudor Sedan
Owner: Jerry Pfitzinger—Kirkwood, Missouri

Old style flames give the 1950 Ford Tudor a classic look reminiscent of the days when customs were the rage.

Jerry and his son Scott demonstrate the technique for removing the Carson style top for those warm St. Louis summer night cruises.

"Hot licks" theme was carried over to the interior where matching flames frame the dash clock. Seen below is the custom radio installation, accessory gauges and the load leveler control.

The old adage of "shoemaker's kids don't have shoes," and "the doctor's kids are always sick" doesn't fit Jerry Pfitzinger of Kirkwood, Missouri. Jerry owns Jerry and Son Automotive, in St. Louis and grew up loving hot rods and customs. He was always building something, or working on his buddies' cars, so it only seemed natural that he go into business for himself. It also followed that he should continue to own and build sharp cars for his own use.

Jerry remembered the little car magazines of the Fifties and how many of them featured chopped Carson top shoebox Fords. He couldn't locate a convertible, but he wouldn't let that stop him from obtaining his goal. A friend told him about a '50 Ford Tudor Sedan he could get reasonable and the "light bulb" went on. He could cut the top off and build his own Carson top. After all, how difficult could it be, he had read numerous articles by people like George Barris, the guys at Gaylord Top Shop and others in *HONK*, CAR CRAFT and ROD & CUSTOM. He would dig out the old books and do it himself.

He contacted his good friend, Jim Stallman and they started in. The easy part was cutting the top off...piece of cake. The real work began after they chopped the windshield 6" and finished the rear package tray off. Using convertible top bows as a guide, they cut and fit their own framework for the soon to be added padded top. Many hours of cutting and fitting, removing cutting and refitting went into this project but in the end...the owner handled project was a success.

The body was already straight, so it was on to other areas. Jerry frenched the headlights, using '53 Chevy truck headlight rings. He turned them sideways and added peaks for a very unique effect. He has always liked the front end on a '50 Ford, so he decided to leave that stock for identification purposes. Out back he frenched and tunneled the stock taillights into the stock position. Since the trunk lid and door handles had already been removed by the previous owner, Jerry felt it was time to take the car over to Donovan Auto Body in St. Louis and have them prep it for paint.

To keep with the Fifties look Jerry wanted, black was decided on for the body color. Donavan used Ditzler black urethane with white, yellow, orange

Side view shows off the Ford's low stance complemented by Olds 88 flipper hubcaps on red rims, long lake pipes and cut down Merc skirts.

Rear of car carries the patented Rotten Ron striping theme showing the name Passion Pit and giving recognition to Donovan Auto Body.

Custom tonneau covers the back seat and matches up with the convertible boot designed by Steve Hasenbeck of Peacock Upholstery. Red and white theme is carried throughout the interior.

Here we see Jerry Pfitzinger (R) and his friend, Jim Stallman (L), building the wood framework for the padded top from scratch.

and red shading to create the "hot licks" look, finishing off the tips in blue. Once the car was rubbed out to perfection, it was ready for striping. Each flame was outlined in contrasting blue or white lines while the rest of the car received bright red accent lines by Rotten Ron. Out back Ron did what he is known best for by creating another Fifties trunk design and the name Passion Pit.

From here the vehicle went to Peacock Upholstery, also of St. Louis, where Steve Hasenbeck worked his magic on the interior. A Pontiac Catalina electric front seat was installed and everything covered in red and white Naugahyde with dark red carpets. The top frame was padded with 2" of padding and then covered with white convertible top material. Steve also created a custom tonneau cover and convertible boot for those topless cruises.

With this done, Jerry brought the Ford back to the shop for final assembly. He installed a 390 Ford big block and automatic, with an owner mounted shift kit and floor mounted shifter. Cadillac load levelers were placed at all four corners to smooth out the ride of the cut coils in the front and the 3" blocks in the rear. Working lakes pipes were hooked up to the Walker exhaust system, while 4" stainless exhaust tips direct the gas out the rear.

The dash was treated to a custom radio section that holds the am/fm stereo cassette player and the accessory gauges. To complete the total rework, all new wiring was added, updating the system to 12-volts. A central fuse panel makes for easy access.

Some things change some do not...thirty years later, Jerry is still working on his buddies' cars and driving a neat ride to work every day. Who ever said that change is inevitable failed to mention that, in some cases, you can delay change until you are ready for it. As far as Jerry is concerned, this is the life he loves...why change!

by Gene Winfield

DREAM FIFTY

Vehicle: 1950 Ford Convertible
Owners: Ed and Dana Guffey — Kansas City, Missouri

When Ed Guffey wanted a custom that would be the answer to a dream, he called on the able talents of Doug Thompson of Kansas City, and this is the result. Appropriately, this shoebox is named Just A Dream.

Starting with a 1950 Ford convertible, Doug chopped the top 4 inches and added a lift-off padded version. In the process of the chop, the windshield was butted and glued at the center. To get everything down to the correct ride height, the frame was C'd and Z'd. Then Al Bradshaw was called on to build the powertrain, consisting of a 302 Ford small block engine with 3/4 race cam and twin 4-barrel carburetors. Gearing is handled by a C4 automatic.

The hood's front lip was accentuated and leads into a '50 Mercury grille opening, where a '54 Chrysler grille was mated to '55 Pontiac bumper ends. Headlights were frenched, with scoops above, and the fenders include "shark gill" portholes with hidden lights. The louvered hood has rounded rear corners and includes a bullnose chrome center strip. Original slab sides of the body are now sculptured in the manner of a '50 Ford Crestliner, sweeping down to handmade '53 fender-length skirts with Merc teeth. Everything is neatly accented with '50 Buick side trim. Lakes pipes are molded into the rocker panels.

Taillights are from a '56 Olds 98, turned sideways. Trunk and doors are operated by electric solenoids. Bumpers are '51 Ford items, with all holes filled, and 1958 Chevrolet license plate shrouds were molded in prior to rechroming.

Bob Sipes handled the full custom upholstery job. An air conditioning unit is molded into the center of the dash, and the radio was moved inside the glove compartment. Dash knobs are handmade. Hubcaps are Dodge Lancer, with spoke additions. A custom lacquer paint job, in a combination of Midnight Blue Pearl with Sky Blue Pearl for the side inserts, is enough to knock your bobby sox off.

Both front and rear bumpers are custom made by mating '51 Ford bumper with '58 Chevy license plate shrouds molded in. All holes were filled and custom work done before rechroming.

A dream custom right out of the Fifties, this shoebox is dressed to kill, in its Midnight Blue Pearl paint and padded lift-off convertible top. A 4-inch top chop brought the lid down low, and the windshield was V-butted and glued.

Left-Taillights are from a '56 Oldsmobile 98, turned sideways. The trunk and doors were shaved and made to operate by electric solenoid. Note exhaust system exiting through '58 Chevy rear license plate shroud.

Right-Interior of trunk was upholstered to coordinate with theme of interior. Spare tire cover says it all.

Left-Handmade '53 Ford fender-length skirts feature a trio of Merc teeth and hide the rear tires and wheels completely.

Six rows of louvers grace the hood, and a pair of spotlights deliver the crowning touch. The '50 Mercury grille cavity is filled with a '54 Chrysler bar mated to '55 Pontiac bumper ends.

Above-Side trim is from a 1950 Buick, and lakes pipes add that Fifties custom look. Note the three "shark gill" portholes on the front fender. They conceal hidden lights.

Right-Lest anyone think that this car is all show and no go, here's a photo of what is under the hood. A built 302 with 3/4 race cam and twin 4-barrels will move this shoebox like a hot rodder's dream.

by John Lee

MONY PIT

Vehicle: 1951 Ford Crestliner
Owner: Larry DeWitt — Wichita, Kansas

Ford was caught off base, so to speak, by the handsome Bel Air 2-door hardtop which joined the Chevrolet lineup for 1950. Not only did Ford not have anything comparable, it also didn't have the funds to develop a new body style, after all that had been spent to redesign the entire line for 1949, a style that was to last the full three-year cycle.

Ford's response to the Bel Air was the Crestliner, a dolled-up tudor sedan that debuted in the spring of 1950. Gordon M. Buehrig, a member of the Ford design staff, had gained fame in the 1930s for his imaginative designs for Auburn, Cord and Duesenberg. It was by his inspiration that the Crestliner got its distinctive two-tone paint scheme. It was a recreation of the "LeBaron sweep" that had appeared on some Duesenbergs and other custom-bodied cars of the 1930s.

Ford's version employed a narrow stainless steel molding that rose from the center grille bar extension in front of the front wheel opening, traced rearward along the side, then swept back forward in a smooth arc ahead of the rear wheel opening. This became a demarcation line between new, bright color combinations such as chartreuse or red, two-toned with black.

To give the Crestliner a sort-of convertible look, the roof was covered with a textured vinyl, matching the color sweep. Aside from the special trim, the sides were devoid of bright moldings. Other standard features were 3-ribbed fender skirts and special full disc wheel covers with blackened spots resembling air vents, a throw-back to the Buehrig-designed '36 Cord. Interiors were upholstered with luxurious new fabrics, color keyed to the exterior.

Although introduced late in the season, and priced $200 above a Custom tudor sedan, the Crestliner sold reasonably well. A total of 17,601 units left the factory before the model change-over.

Buehrig, as manager of Ford's Body Development Studio, was assigned the task of developing a pillarless hardtop for the 1951 Ford line, that could be built from the convertible's tooling. But it wasn't ready at new model introduction time, so the Crestliner was updated to the new styling to fill in. The major change from 1950 was retention of the Custom's stainless rub strip that now ran in a straight line from behind the front wheel opening, back along the side, and all the way around the rear, underneath the deck lid. The Crestliner trim, therefore, joined the other trim aft of the door opening, and the two-tone treatment continued below the trim line clear around the back.

Once the Victoria hardtop appeared, it naturally stole the Crestliner's market and went on to rack up sales of 110,000. It beat Chevy's Bel Air by 7000 units, in the process. Crestliner sales, meanwhile, slacked off to 8703 for the year. It would not return as a special model, although the name Crestliner was adopted for Ford's top series for 1952.

Larry DeWitt of Wichita, Kansas, bought a green and black '51 Crestliner in 1955. He repainted it black and white, installed a set of duals and courted his wife-to-be, Mary Lou, in it while he was stationed with the Air Force in Fort Worth, Texas.

A couple of years back, he set out to find another Crestliner and replicate his courtin' car. A bronze and brown one was located, with a Ford-O-Matic, which, though not exactly the stick-and-overdrive of the original, requires less work to drive. Warren Brandes at Brand Ford Garage took care of restoring the mechanicals, including a pair of Fenton exhaust headers.

Larry's personalized restoration called for a black metallic and white pearl finish, sprayed by Daryl Kiddricks, and two-tone gray velour upholstery by the Morgan-Bulleigh Company. Lewis Street Glass installed smoked glass all around, and Abe's Plating took care of rechroming the many trim parts.

Larry's 'Liner captured a second place trophy in Postwar Hardtop its first time out. But mostly the '51 Crestliner is for him and Mary Lou to relive those golden days of youth.

Larry's Crestliner is one of only 245 registered with the 1949-'51 Ford Registry. For more information, contact Mike McCarville, Box 30647, Midwest City, OK 73140; (405) 737-6021.

Beautifully restored '51 Crestliner looks like it just stepped out of yesterday. Black metallic and pearl white finish was applied by Dale Kiddricks, and Abe's Plating took care of rechroming the many trim parts.

Interior was brought back to original condition, but for the two-tone gray velour upholstery, stitched by the Morgan-Bulleigh Company.

Under the hood is a fully restored flathead powerplant, complete with a pair of Fenton headers. Backed up by a Ford-O-Matic, this car is easier to drive than Larry's original stick-and-overdrive Crestliner.

All the trim is original, and the fender skirts conceal a set of wide whites, for the correct atmosphere of the early '50s.

Part of the original Crestliner equipment was the chromed splash guard just behind the front wheel openings. These items have all been rechromed to a factory sheen.

An experienced spare tire shows miles of wear, and the narrow whitewall is evidence that, although this is a trophy winner, it is not strictly a show car only.

by Ron Ceridono

ELVIS '51

ROCK AND ROLL CLASSIC

Vehicle: 1951 Ford Club Coupe
Owners: Bill and Sherry McGarity—Rio Linda, California

Front of the '51 has been cleaned up considerably by nosing the hood, frenching the lights and swapping grilles. Note also the Buick portholes in the top of the hood. Bill is not a professional body man, just a hobbyist, yet his workmanship is flawless.

When Bill and Sherry McGarity brought this '51 Ford home, it looked pretty cherry. Then they found a little plastic filler, no big deal. Then they found acres of filler, the deal got a little bigger. Bill decided to make the Ford straight, and add some custom features at the same time. Eleven months and four days later, it was on the road.

Starting up front, a '54 Chevy grille with extra teeth replaced the stocker. The headlights were frenched, the hood has been nosed and fitted with Buick portholes.

To keep the smooth look going, bumper guards were removed and the bolt holes were welded shut. The bumpers were then smoothed up and replated.

Doors and trunk are solenoid operated, rear lighting is by frenched Cad lenses. The right rear fender has received dual sunken antennas. Unique hubcaps were made from Plymouth cones and Oldsmobile pieces.

Power for the '51 is provided by a stock flathead and drivetrain. A combination that has proved reliable enough to haul Bill and Sherry all over the west coast.

The McGaritys are Elvis fans and they chose to show it with some details on the car. Hand made musical notes appear on the door panels, lighted "Elvis '51" door sills and a mural on the continental kit complete the theme.

All the body work on the car, including the House of Color candy red paint, was performed by Bill, with help from Lar Hicks, Randy Choy, and an understanding, supportive wife.

What's next? If they can stop driving it long enough, a full boogie flathead.

The Club Coupe has been lowered 3-1/2 inches all the way around. Air shocks and an onboard compressor are used to increase clearances when Bill and Sherry hit the road. Pontiac side trim replaces the original. Fender skirts, dual sunken antennas, a continental kit, and lakes pipes add to the custom flavor.

No self-respecting '50s custom owner would leave his pedals uncovered. Red and white interior is period perfect. Teardrop and dice knobs are featured, along with fuzzy dice and a "neckers" knob.

Musical notes on the door panels were handmade in tribute to Elvis, and the rest of the car's theme shows Bill and Sherry's devotion the the King of Rock 'n Roll.

'50s style red and white interior trim carries over to the trunk. The gas filler door has been smoothed off, the filler has been moved inside. Frenched Cadillac lenses perform taillight duty.

Unique hubcaps were handmade by combining Plymouth cones and Oldsmobile pieces. Wide whites are the right accent for this period custom.

by Jim Genty

MISS VICKY

Vehicle: 1951 Ford Victoria
Owners: Jim and Mary Ann Genty—Brea, California

Late Fifties/early Sixties look is enhanced by the two-tone scallops reminiscent of the great Larry Watson.

Blue dot taillights, no bumper guards and the scalloped license plate hood cry out the "lost in the '50s" theme. Note the rare dealer installed bumper hitch.

Classic lettering by Herb Martinez gives the Vicky its own identity, reminiscent of a time when everyone gave their cars names. KINGSMEN plaque can be seen in the rear package tray.

Smooth bumper theme is carried over at the front. Notice how smoothly the scallops flow back on the hood and the classic look of the Plymouth hub caps.

Did you ever think how one goes about establishing a corporate identity? Ford has their blue oval, Chevy the bowtie, McDonalds the golden arches. This was a problem I was faced with after deciding to establish a mail order business, selling t-shirts, caps, jackets, radiators and custom suspension parts to '49-'51 Ford owners.

At the same time I was wrestling with this problem, I was trying to find a shoe box Ford to build and drive to my old car club's 32nd reunion in Denver, Colorado. A surprise phone call from a good friend of mine and Kingsmen alumni, Joe Konkright, solved both problems.

"I just bought a 1951 Ford Victoria and I am already two cars over my limit. Are you interested?" Never one to pass up an opportunity, I jumped on the next available flight to Denver to check out the car.

When I got there I was more than satisfied with

what I found. While Ford made over 110,000 of the Victoria model in 1951, they seem to have been swallowed up by every crusher in the country.

A brief orientation of the history of the car by Joe informed me that it had been built by Cliff Adenbrook of Littleton, Colorado. He purchased the car out of New Mexico from the original owner in 1988 and immediately set about making it into a cruiser. He yanked the original flathead and three-speed trans out, and replaced them with a 1975 Ford 351 Windsor from a Ford F-100. He exchanged the C-6 automatic for the smaller C-4 to get better clearance at the floorboard. Out back he installed a 1966 Mustang 8-3/4 rearend with the stock springs from the Ford. This is what sold me on the car, late model reliability for lots of trouble free miles. The engine was detailed out with braided steel line, Ford Motor Sports valve covers and a finned aluminum air cleaner from a 390/401 Ford Galaxie 500.

At this point Cliff was ready to have the car painted. He took it into Fall Guys in Denver, Colorado to have the body massaged in shape and a custom paint job applied. They chose '85 Thunderbird light and dark taupe, light and dark brown to you old timers. While the car was torn apart, all the chrome was sent over to Alert Chrome (also in Denver) for a complete redo. Once the paint was rubbed out and the chrome put back on, the Ford moved on to Mark Ford of Custom Comics for one of his patented striping jobs. All of the scallops were out lined in a contrasting shade of brown.

Now the car was ready for a new interior. A & D Striping (I know, I know...strange name for an upholstery shop) pulled the old interior out and added white pleated Naugahyde to both seats and the door panels along with a new white head liner. New black carpets replaced the old rubber ones. This same theme was carried over to the trunk where all new panels were made and covered with the same white pleated design used in the interior. Black nylon carpet completes the trunk compartment.

Cliff chose dummy Appleton spots, and long chrome lakes pipes from Night Prowlers in Kansas City, Missouri to accent the late Fifties, early Sixties look of the Victoria. To complement the 14" steel wheels and white wall tires, he chose late model Plymouth Satellite hub caps to set the look off.

The two things he did not do was to lower the car and change the radiator. Since I sold both of these items and since this was going to be my new corporate logo, I immediately set about installing both of these items when I got back to California.

I pulled the stock coils out and installed new custom JAMCO Engineering 3" dropped coils along with our shorter A-Frame snubbers and new sway bar bushings. Out back I installed 3" lowering blocks and our Cruiser gas shocks. Now it sat right, and still rode like a dream. No more roll over tendencies when I went around the corners.

Engine compartment is stuffed full with 351 cubic inches of late model Ford power, accented by braided steel lines, and finned aluminum air cleaner.

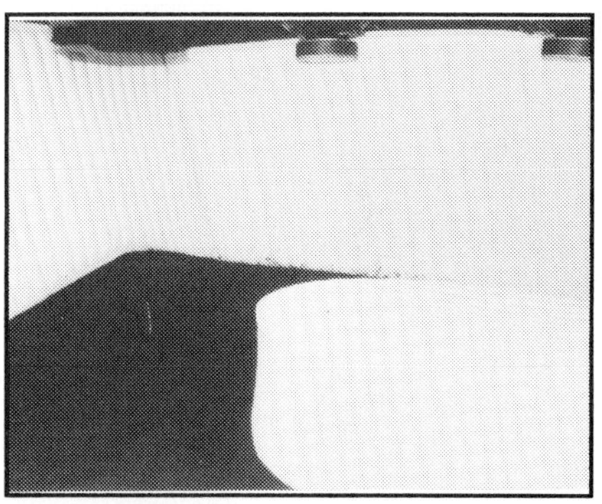

All white interior is broken only by the painted upper window molding and the chrome door frames. Black carpet contrasts sharply.

Vintage window water cooler is a real conversation piece with its matching paint and Vintage Air lettering.

The next step was to pull the radiator and install a JAMCO Engineering 4-core unit. These units are actually smaller than stock but have 300% more cooling area. Perfect for cruising.

Since purchasing the Vicky a year ago, we have put over 18,000 trouble free miles on her, captured Best In Show/Custom in Las Vegas at the First National Convention for '49-'51 Ford Owners and picked up numerous People's Choice awards. It may never be another bowtie logo or Golden Arches, but it certainly has become synonymous with J&M Enterprises and JAMCO Engineering. See, creating a corporate logo is not so hard and sometimes it's even fun!

by Rich Johnson

SHOEBOX FORD/MERC

FRAME-OFF FIFTY THREE

Vehicle: 1953 Ford Convertible
Owner: Ken Mathias — Lakewood, Colorado

This is the way a frame-off project looks while all the work is being done. From the repaired spots, it's easy to see all the areas of floor sheetmetal that needed serious attention.

Ken Mathias is one fortunate man. He's a guy who still owns his very first car — the car he had in high school, the car of youthful memories. But today, his first ride looks far better than it ever did back in the good old days. Here's how it all happened.

Back in November of 1960, he saw an ad in the local hometown newspaper for a 1953 Ford convertible that was for sale. The following Saturday, he went to see the car, but failed to make a decision fast enough, and the car slipped away. Price was $200.

As if heaven was smiling on Ken, he saw another ad in the local newspaper the following March — a 1953 Ford convertible for sale. He went to see the car, and it turned out to be the very same convertible. The price had gone up to $225, but a new set of tires and a fresh windshield had been added. This time, Ken didn't hesitate. He bought it and took it home.

Between 1961 and 1966, Ken worked on the car and drove it to high school, and later to college. The only work that was done was just enough to keep it going, because funds were severely limited (a familiar story to kids that age). In '66, he purchased a Corvair and put the Ford in storage in a Kansas barn, because he figured it was only worth about 50 bucks.

It was 1982 before the convertible saw the light of day again. When it came out of storage, the tires still had 15 pounds of air in them. Ken filled the tires to full pressure, hooked up a tow bar, dropped the driveshaft, and towed the car 400 miles to his home in Lakewood, Colorado. Renovation started immediately thereafter.

Originally, the car had been mildly customized, with frenched headlights, a shaved hood and deck, and changed side trim. Ken decide the car would best suit his personality if it was restored to original condition in every way. So, he began to de-customize it.

Over the next several years, more than 2000 hours of work went into the restoration. Major rust damage

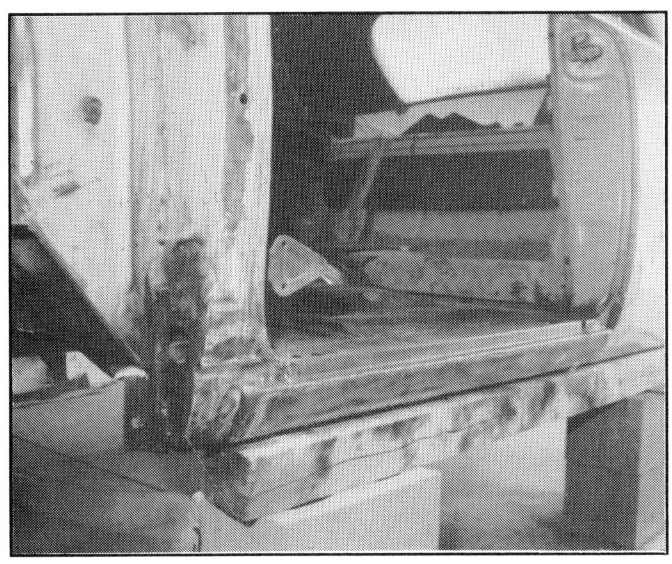

A replacement rocker panel was needed because of the amount of rust damage that had occurred over the years. What couldn't be simply repaired was replaced with NOS pieces.

With body work completed, Ken prepped the car for paint. Obviously this isn't a hot rod, but Ken did all this work right in his garage and driveway, which is absolutely the spirit of hot rodding.

had taken its toll, and even before the car went into storage it had been in poor condition. So there was plenty of work to be done. Everything came off, right down to the frame, so nothing in need of attention would be missed. Much of the sheetmetal required repair or replacement. NOS front fenders were installed, and a wrecking yard hood and grille found a new home. A '54 deck lid was located, filled and redrilled to accept '53 trim. All the hinges and latches were either replaced with NOS pieces, or refurbished to deliver original standards of fit and function. Even the original top mechanism was completely disassembled, overhauled and repainted in OEM color. Pivot pins were rechromed and a reproduction

The convertible top mechanism was fully disassembled and restored to original working condition. Even the pivot pins were rechromed. Roll-up window mechanisms were removed, restored and replaced.

The dashboard was completely disassembled and everything behind it brought up to like-new standards. Here we get a peek at the heater and windshield wiper mechanisms, which were fully restored.

stainless steel latch handle installed.

All the mechanicals were refurbished, including new seals, bearings, gaskets, and anything else that was necessary to make the car like new again. The original flathead block had cracked, so nearly 20 used engines were disassembled and Magnafluxed to locate a good one. When all was said and done, the engine going back into Ken's convertible was equipped with balanced Mercury internals and Offenhauser 8:1

heads. A Mallory Magspark ignition (bought for $5 in 1962) was used, but a new coil was installed for reliability. He rebuilt the 3-speed overdrive transmission, balanced the driveshaft and installed new U-joints.

On the inside, the dashboard was completely disassembled, restored and reassembled. The seats came out, door panels came off, and window mechanisms were removed and repaired to like-new condition.

Ken performed all the labor to restore this beautiful '53 Ford as far as mechanical, wiring, body and paint work was concerned. The only work he farmed out was machining and balancing of the engine, straightening and plating some trim pieces, and interior upholstery work. Everything else, he did himself. Believe it or not, he even painted the car in his driveway.

It is quite obvious that Ken Mathias is not only one lucky man, but a heck of a talented guy, as well.

With the dashboard back in place, the pedals hung, and the steering wheel positioned, Ken's convertible is starting to take shape. Photos like this make us drool and appreciate all the work involved in a frame-off restoration.

What a great looking firewall! Looks as if we stepped back in time to '53 and paid a visit to the Ford factory. Wiring has all been replaced, and every component is as new.

As part of the underhood restoration, these inner front fender panels were installed. Note the battery box, horns, heater motor, fresh air intake, and other components.

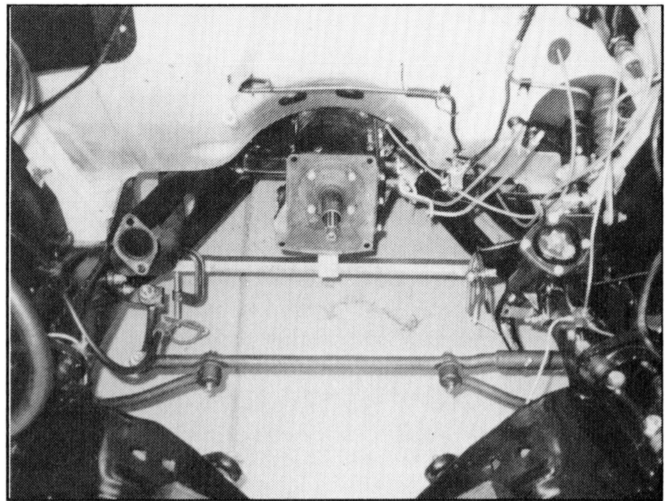

Engine compartment is ready to receive a freshly built flathead. All the steering components, exhaust system, plumbing and wiring were redone to factory spec. Even the 3-speed overdrive transmission was overhauled by Ken.

The original flathead block had cracked, and 20 wrecking yard blocks were checked by Magnaflux before a suitable block was found for rebuilding. The finished motor was filled with balanced Mercury parts, and 8:1 Offenhauser heads installed.

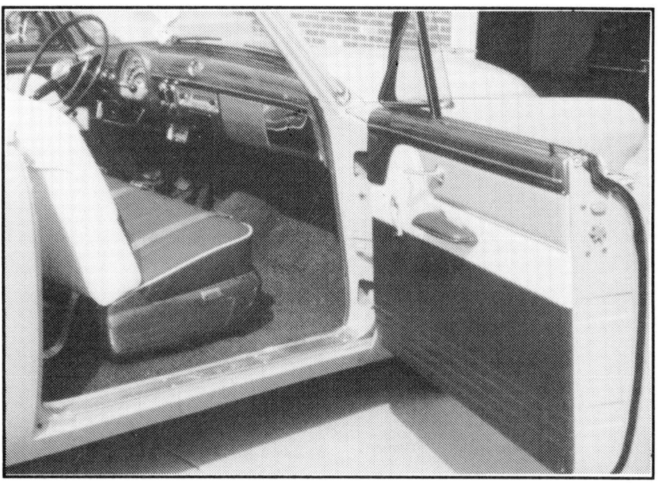

The interior looks factory fresh. All the upholstery, including seats and door panels, was reproduced to exact original color and pattern. The floor was carpeted in dark green.

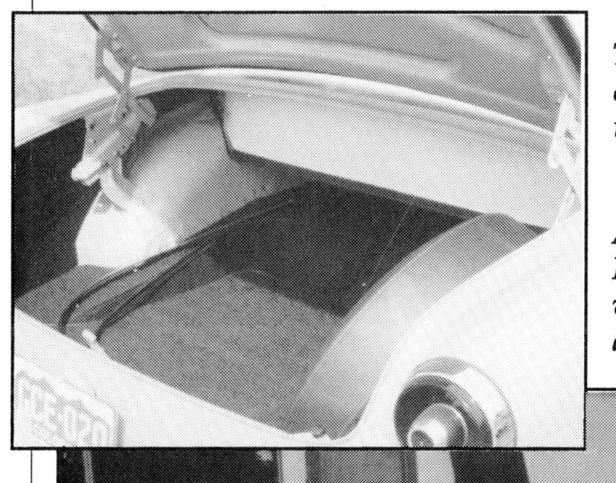

Trunk carpeting matches that used in the passenger compartment, and the spare tire has a vinyl cover to match the upholstery material for the seats.

A fitting garage for such a fine example of a restored '53 Ford convertible. From every angle, this is a superb looking vehicle, and Ken can be rightfully proud of the quantity and quality of work he did.

by John Schumont

WILD CHERRY

All I ever wanted was a car with spotlights and shaved door handles.

Vehicle: 1953 Ford Convertible
Owner: John Schumont — Plymouth, Michigan

It seems like everyone remembers a certain type of car when they were growing up. In my case, I remember the customs of the '50s and '60s. Having always wanted a car with spotlights and shaved door handles, I located a rusted-out stock 1953 Ford convertible in Indiana and trailered my prize home.

First order of business was to eliminate some glass area, so I chopped 2-1/2 inches out of the top, while still retaining full folding action. The ride height was way too high, but a Firebird subframe brought my car 5 inches closer to the ground.

The front end was smoothed out by frenching the headlights, removing all trim, molding in a new grille shell, rounding the hood corners, and that airplane had to go! A modified '55 Ford grille now fills the cavity. While I was at it, bumper bolts, front and rear were shaved.

Most custom cars have Lakes pipes and mine has them too, except with a twist. I made my own pipes with the exhaust opening split in half and frenched in the area where rocker panels once used to be. Reversed louvers fade into the hand-formed scoops that start at the front of the rear fender bulge. At night, a blue haze glows from behind these louvers and the louvers below the deck lid.

I extended the rear fenders slightly and molded them around '56 Olds taillights. Naturally, the deck lid was shaved, and a frenched license plate housing finished off the rear of the car. On the left rear fender is a sunken antenna.

The powertrain consists of a 350 Pontiac engine with a hot rod cam, connected to a Corvette 4-speed which spins a homemade driveshaft to a Maverick rear axle.

The interior consists of a 6-way power Chrysler front seat, covered with white vinyl and off-white velvet. The rear seat is stock and is covered with the same material, along with the doors and side panels. Steering chores are handled by a Chevy Impala steering wheel, mounted on a Pontiac column. Red nylon carpeting is complimented by the interior trim and gold striping on the dash.

A six-inch drop was achieved in the rear with blocks and de-arched springs. This requires a stepped frame and raised driveshaft tunnel.

After I achieved the proper silhouette, I applied 17 coats of red garnet lacquer to the exterior, topped off with white pearl scallops. Gold pinstriping by Paul Hatton is used for accent.

I built the entire car, with the exception of the interior. After four years of work, I've brought back a little bit of my boyhood. Oh, by the way, my car has spotlights and shaved door handles.

In the beginning, there was rust. Actually, the original car is not in bad condition, considering that it came from Indiana.

One of the first things done was chopping the top 2-1/2 inches. This was done in a manner that allowed the full collapse and extension of the folding convertible top mechanism.

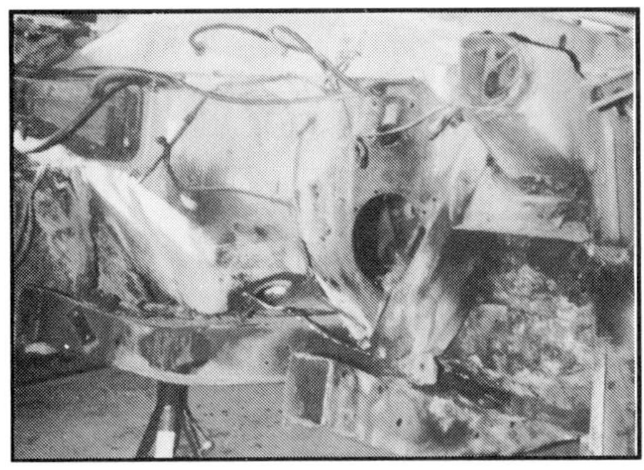

Above-The front portion of the stock frame was cut off near the firewall in order to permit the grafting-on of a '68 Firebird subframe.

With the Firebird subframe welded in place, the custom Ford now has independent front suspension, and more modern front drum brakes. The rear portion of the frame was stepped to bring the car 3 inches closer to the ground, and the Firebird front end took care of that up front. Lowering blocks were employed to bring the car down another 3 inches.

Left-In a semi-completed stage, the old Ford is starting to take on a new profile. Headlights were frenched and the hood was filled with louvers. Skirts are in place, but the door corners have not yet been rounded. There's still plenty to do.

Below-By building real scoops and adding reversed louvers, the Ford rear fenders now become functional as they channel fresh air down around the brake area.

Left-Trunk corners were rounded, louvers were added to the panel directly below the deck lid, and the licence plate area was boxed and sunken. Notice that the tail-lights have been extended.

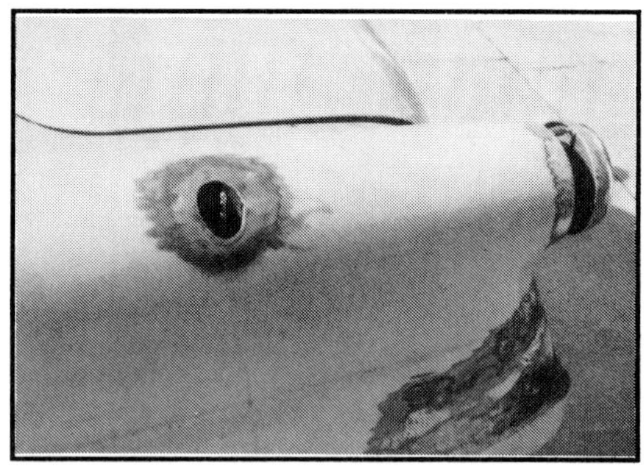

Above-On the driver's side rear corner, an important custom feature was built in — a sunken antenna.

Left-Impala steering wheel sits atop a Pontiac column. Transmission is shifted by Hurst 4-speed shifter. Note the use of louvers in the glove box, steering column support and door trim.

by Jim Clark

GREEN FLAMINGO

Vehicle: 1949 Mercury 4-Door Sedan/Convertible
Owners: Tony and Joyce Besase — Virginia Beach, Virginia

In the beginning, the Green Flamingo was just a mundane rust colored 4-door sedan, but Tony and Joyce had a vision of better things to come. Through the conversion, they retained all 4 doors, and left the rear ones suicide style.

First order of business was to get rid of the top. Here the car is at Dick Dean's shop, where the master's touch was applied to top removal and chopping of the windshield.

Rear quarters have flush-mounted, flared skirts and scoop built into front edge of raised panel. Rear doors still function in normal suicide fashion. Teeth in scoop are '53 Merc grille items.

Custom cars in the '50s were built from a wide variety of makes and models. Outstanding show stoppers were constructed from among these, with a few standing out above the rest. Super Stars of these were, and still are, the '49 thru '51 Mercurys.

Their popularity makes them prized subjects to base a custom project on, but only if they are the right model, namely the 2-door sedan or convertible. The popularity of these models has made them scarce and expensive.

Some enterprising custom car enthusiasts have come up with a less expensive alternative, the 4-door sedan. In its unaltered form, the 4-door is not that desirable, but slice the roof off, add a Carson top and you have a very unique cruiser. That was the course chosen by Joyce and Tony Besase of Virginia Beach, Virginia.

After a few years of heavy involvement with aircraft and boating activities, they have returned to their first love, cars. They obtained a prime condition, box-stock '49 Mercury 4-door. Then had the hood corners rounded, head and taillights frenched, emblems and door handles shaved by Cecil Profit of Portsmouth, Virginia. Next it was shipped off to southern California for the major modifications.

Grille bar was fashioned from '54 Ford center, with spinners added at ends. Small accessory lights above bar serve as turn signal/parking lights.

There, the top was cut off and the windshield chopped. A molded fiberglass top was fitted and covered with standard convertible top fabric. The removable top fastens in place with over-center style latches.

Fully functional power windows, front and rear,

fill the gaps on each side. Parts and pieces from early '52 through '54 Ford hardtops were combined to create the vent and side windows. Chrome trim atop the doors is from the same source. Power window motors, switches and mechanisms are all late GM components. Rear window in the Carson top is a standard convertible unit, available from aftermarket top suppliers.

Both the front and rear pans behind the bumpers were molded in, and the grille cavity smoothed. Components from a '54 Ford grille fill the opening, topped by a pair of accessory parking/turn signal lights.

A pair of power antennas reside in sunken mounts atop the right rear quarter. Below, the raised panels on each rear quarter have been filled and fitted with flush-mounted flared fender skirts. A scoop was fashioned at the leading edge of each panel and filled with three '53 Mercury grille teeth.

To smooth the lines of the front-to-rear fender transition, a panel was added, taking the hump out of the line at the rear of the front doors. Both front doors have solenoid operated latches, while the rears open manually from the inside only.

Two shades of green were selected, highlighted by contrasting multi-colored graphics on each side. Pinstriping and a stylized flamingo were added on the panel near the antennas. All of the paint work was done at Cypress Auto Body in Cypress, California.

Though the interior looks very elaborate, it was a relatively easy conversion. A complete interior from a '65 Thunderbird was obtained, then upholstered in white vinyl with green fabric inserts. Matching door panels, carpeting and headliner in the Carson top were also fashioned from the same materials. The T-bird center consoles were retained, covering them in the same manner. Up front, the center console extends from the floor up to the bottom of the stock '49 Mercury dash. The stereo and speakers reside in the console. All of the interior work was performed at R&R Custom Upholstery.

Few modifications were made to the drivetrain at this time. Only some refurbishing of the stock flathead V-8 and addition of a dual exhaust system were done, with plans for future engine mods and detailing to come later. Lowering blocks at the rear, and cut coils up front, bring the car down to its ground-hugging stance.

Joyce and Tony built the car to be driven, not just as a display piece. Since its completion, it has appeared in two shows, winning both trophies and acclaim, while being driven a considerable distance to each of them. Our last view of the Merc was of its blue dots heading eastward on a trip from California back to home base in Virginia. A worthy test of any contemporary cruiser.

A fiberglass Carson top was fitted to the open Merc, and was later upholstered inside and out. Dick Dean is the master when it comes to chopping Merc tops, and making the vent windows fit.

Joyce and Tony Besase can soak up the sun in real style aboard their fully functional '49 Merc 4-door open cruiser. Carson top and roll-up power windows make everything tight during foul weather.

Interior is from a '65 T-Bird. After re-upholstering, it fits right into Mercs without alteration.

Dash is stock '49 Merc, painted white. Console is '65 T-Bird, as are front bucket seats.

by Jim Genty

LOW RIDER

Vehicle: 1950 Mercury Business Coupe
Owner: Gene Elgan—St. Louis, Missouri

Classic lines of Gene's Merc are accentuated by the rounded hood corners, V'd and narrowed '53 Chevy grille bar and the frenched stock headlight rings.

Sandalwood beige Naugahyde interior is reminiscent of days gone by when their was room for "you and your baby" and the rest of the gang.

Clean trunk area, the frenched '51 Buick taillights, and his ALKY BURNERS club plaque reflects Gene's fondness for the "good old days."

Have you ever noticed how some people change cars as often as some of us do oil? It seems like every few months they are driving a different car. On the other hand, there are guys like Gene Elgan who keep the same car forever. Gene's philosophy is, "Why change cars when you can just keep making the one you got better!" A noble thought, but then he can back it up in spades!

Gene first acquired his 1950 Mercury coupe back in 1969, which meant he owned the car all through the years of the '70s when you couldn't give a custom car away.

He started off like most of us, driving the car in basically stock condition. But reading all those little car magazines from the Fifties had kindled a spark that lay smoldering, waiting for the time when it could burst into a flame.

Around 1971 primer spots began to show on the old Merc body as he gave it a traditional nose and deck treatment and added fender skirts. A mild

The slanted and rounded door post along with the '55 Buick side trim give the normally rotund Merc a slick image of motion.

Here we see Gene's Low Rider sitting in front of General Grant's farm, near his home.

Slanted top line is evident in this shot showing he V'd front windshield and the small peek mirror used for "backward glances!"

The name Low Rider is from a time when "low and slow" meant custom car. Lettering is the handiwork of St. Louis' Rotten Ron.

Here we see the classic look of wide white walls and '57 Cadillac hub caps. Note frenched antenna and proximity of lakes pipe to ground.

Under construction photo shows car just after the top was cut and the door post slanted forward with rounded corners.

lowering job and he was ready for the first of what was to be many paint schemes. This one was "James Dean Black."

The trusty Merc stayed this way until 1973 when he felt the urge to change colors. This at a time when most people had all but forgotten about customizing cars, let alone '49-'50 Mercurys. This time he opted for 1955 Oldsmobile dark metallic green. He drove it like this until 1976 when he remembered a great color from the Fifties, and repainted it Honduras Maroon.

Satisfied that this was the ultimate color, he drove it like that all through 1976 and then in 1977, he decided that it was time to get serious about customizing the Mercury. He frenched in the stock headlight rings in front, while out back he used '51 Buick lenses. He molded the stock grille shell in and added a shortened and V'd '53 Chevy grille bar with plenty of extra teeth. It was also time to deep six the door handles. With all this done, it was time to cover up the primer spots again. A deep candy grape was selected with silver metallic below the newly added '55 Buick side trim...this is it he thought, "I'll never change it again!"

Guess what, around 1984, after a span of seven long years and many miles, those creative juices started to flow again. After all, custom cars were starting to surface (slowly) again. Of course Gene never knew they had disappeared. He decided it was time for the ultimate complement you could bestow on a Mercury...it was time to chop the top. He cut away 4-1/2" of unwanted metal, slanted the door posts forward while rounding both the door corners and the hood corners in the same process.

A power antenna was frenched into the front fender and the gas tank door removed, putting the filler neck inside the trunk. Satisfied with the new look, he took the car over to Custom Jim Ebenhoh for another two tone paint job. This time Gene selected a dark copper brown over a gold Corvette lacquer. Once the car was rubbed out, he accessorized it with '57 Cadillac hubcaps, full length lake pipes and striping by Rotten Ron of Custom Designs in St. Louis.

Gene says he is satisfied with the look and drives the old Mercury as often as he can. The problem is, he likes Mercurys so much, he bought a second one-owner stocker to drive back and forth to work. The current problem is, does he paint the latest edition, or redo the old Mercury again. After all, pastels are back bigger than ever and so is pink and....

by Ron Ceridono

PRO SLED

Vehicle: 1950 Mercury Coupe
Owner: Keith Novak—Old Bridge, New Jersey

Lines of a chopped Merc are classic. The lid was lowered a whopping 5-1/2 inches by Keith Novak. Weld Wheels are used at both ends as are ribbed bumpers by Mr. Merc.

Keith Novak wanted a sled. But not just any sled. He wanted his to be different. Frenching, nosing, decking, chopping, it's all been done before. What was needed here was modern execution of yesterday's style. Keith set about to create a combination of the old and new, a Pro Sled.

The Pro part of the package began with the installation of a 1973 Chevrolet 454. The Rat motor features a steel crank, 10.5 to 1 compression pistons and an LS-7 cam. Topping things off is a Corvette tri-power induction system. Backing up the big block is a TurboHydramatic 400 with a manual valve body and a 10" A-1 torque converter. This potent combination has propelled the less than svelte Merc to surprising 11.96 elapsed times at 117 mph.

Given the Merc's straight line performance potential, the stock handling and braking capabilities were next to be updated. Up front that problem was addressed by grafting a disc brake, power steering equipped 1971 Chevy Nova clip onto the original frame rails. In the rear, a custom rear frame section uses parallel leafs to mount a Ford 9-inch that has been equipped with 3.90 gears and Strange axles.

Rolling stock consists of 6"x15" Weld Wheels wrapped with G/60-15 rubber up front. In the rear the unmistakable Pro Sled statement is made by

Stock frame was cut off just in front of the firewall and a Nova clip installed. The Merc ran in primer before being painted by Mike Wastog.

Healthy 454 Chevy makes the Merc move. Mr. Gasket scoop hides the Corvette tri-power carburetors and manifold.

Nova clip has been securely fastened to the stock frame rails and X-member. Big block Chev hooks to a Turbo 400 which drives a 9-inch Ford rearend. Headers, part of the performance package, show some dings as a result of the sled inspired ride height.

Ford rearend is suspended on leaf springs from a custom rear subframe. The 9-inch has been equipped with Strange axles to withstand the big block's torque. Mickey Thompson tires supply almost 40" of rubber on the road.

Keith travels light, as most of the trunk space is taken up by the tubs necessary to house the Pro size rear tires.

All the custom tricks have been used on this Merc which has been chopped, frenched, nosed, decked, and smoothed. What sets it apart is the Pro package under the metal work.

A little tire-frying demonstration was put on by Keith to show that the Merc goes as good as it looks.

14"x15" Weld Wheels with monster 18.50x31 Mickey Thompson tires.

Once the Pro part of the package had been taken care of, it was time to get the sled part squared away. Keith determined a 5-1/2" top chop ought to make the proper custom statement so he whacked the roof himself. Up front a whole load of sheetmetal modifications have been made. The hood corners have been rounded, scoops formed on the sides and 164 louveres punched. In the best custom tradition, the headlights have been frenched and a '57 Corvette grille now resides between Chrysler Cordoba parking lights. Bumpers on both ends were supplied by Mr. Merc.

Custom made door buttons replace the stock handles while around back frenched taillights flank a smoothed deck lid. When the metal work on the Merc was done, Mike Wastog covered it with 1989 Cadillac Allante maroon with a gun metal gray stripe. Upholstery was handled by Pete Barbarn.

Keith Novak wanted a Merc that would stand out in the crowd. His Pro Sled not only stands out in the crowd, it'll outrun most of it too.

by John Lee

MAGIC VALLEY MERC

Vehicle: 1950 Mercury
Owner: Carolyn Kissell — Burley, Idaho

It all started when Jim Kissell was a teenager and got his first car, a 1949 Chevrolet. That was followed by a succession of machines: 1959 El Camino, '50 Ford truck, '55 Crown Victoria, custom '51 Chevy, '23 T roadster, and '63 Chevy convertible. (A 1940 Nash Lafayette business coupe is planned for later.) But for now, all the hauling to rod and custom runs is via a 1950 Mercury. And a fine hauler it is.

This particular car, titled D Merc, was started by Carolyn's brother Bo Huff, when Bo was living in Fayetteville, Arkansas. When he moved to Utah to open a custom body/paint shop, Bo brought the Merc with him, and that's where Jim bought it for Carolyn. As always, some minor changes and work were scheduled, which Carolyn says was a mistake. When the mechanic drove the car home, he drove it *through* the Kissell's garage! Back to brother Bo for a repaint.

We wish we could show you this beauty in living color. Covered with Black Cherry Muscadine (House of Colors urethane), the surface pops with bright highlights in sunshine, and takes on a deep hue under subdued light.

Active in the Magic Valley Early Iron Club and the Kustom Kemps of America, the Kissells are rightfully proud of a car that has placed first in ISCA Radical Custom competition, been judged outstanding custom at many shows, and owns plenty of attention as People's Choice.

Johnny Waldon of Fayetteville, Arkansas did the interior, but the trunk was stitched up by A-1 Upholstery in Burley, Idaho. It follows the theme set for the interior.

The combination traditional/modern interior of gray velour uses a 2-inch tufted diamond design on Chrysler Imperial seats. Carpet is light gray plush to enhance the black lacquer dash, now holding Stewart Warner instruments. Floor shifter feeds commands to a TH350 automatic with shift kit. Locking steering wheel is Mercury.

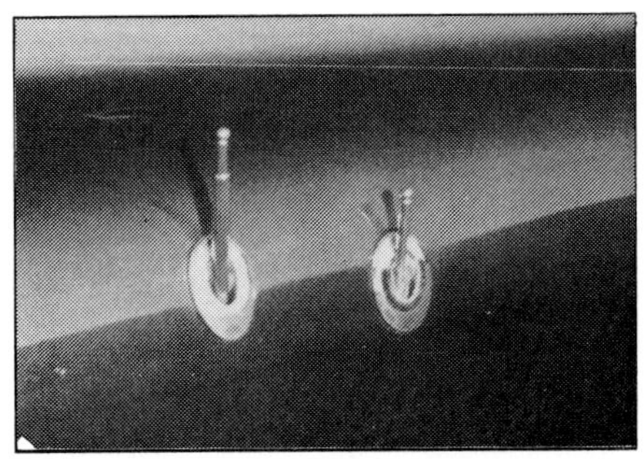

Twin flush-mount antennas in the left door cap send signals to an Alpine stereo unit with booster and four speakers.

Showing expertise he is known for, Bo Huff made the quarter windows operational, a definite plus for air circulation.

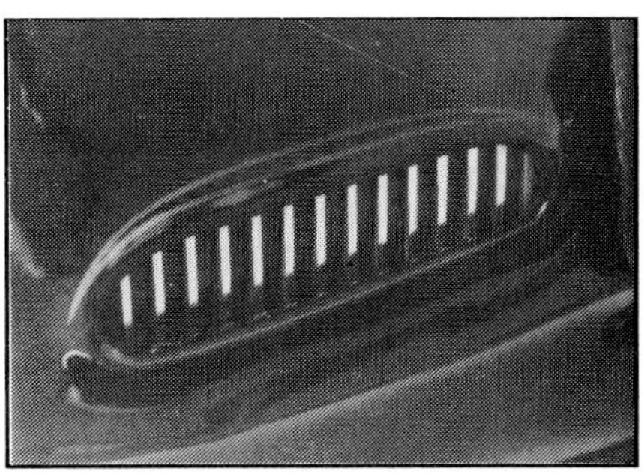

Taillights were frenched and given a slight tunnel effect that was carried through from the headlights.

Headlights are tunneled six inches, chrome rings added along with lamp rock guards. Pin striping was done by Jerry Crum of Grand Junction, Colorado.

The 400-inch Chevy small block includes an RV cam, Edelbrock valve covers, Holley manifold and 750-cfm carburetor, as well as Mallory ignition. All plumbing is stainless, and most removable items in the engine compartment have been chrome plated.

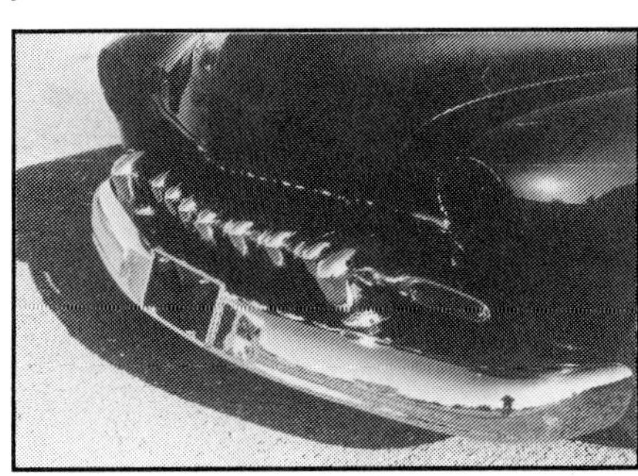

Front and rear bumpers have been smoothed, hood corners rounded, and a '56 DeSoto grille installed in a rolled opening.

An electric fan and shroud are positioned ahead of a 5-row Dodge radiator, which has proven more than adequate for cooling the Chevy engine.

Huff chose to go with the backlight from a '41 Ford, which gives a very pleasing effect in the nearly flat roofline. An interesting departure from current customizing practice is to run without skirts. In the old days, that was a sure giveaway that the custom was running a hopped-up engine.

by Gene Winfield

THE SNAKE

Vehicle: 1951 Mercury
Owner: Gene Duvall — Fort Wayne, Indiana

Gene Duvall named his cruiser the Snake, for obvious reasons. You can't get any lower than a snake's belly, and when this Merc is in the weeds, there is only 1/4" of ground clearance! Low enough to do any serpent proud.

But being low doesn't stop there. The top was chopped 5-1/2 inches and a '50 Merc rear window was installed, bringing the profile way down. Besides being low, the Snake has undergone a bundle of other custom tactics. Headlights are rectangular units out of a Ford, and parking lights are 1980 Mustang, positioned beside the 1955 Oldsmobile grille bar. Front and rear pans were rolled and molded to the bumpers and fenders, and the front wheel openings have been flared, as were the rocker panels. Hood corners were rounded, and six rows of louvers help ventilate the engine compartment, getting rid of heat generated by the 1979 Chevy 350 engine. Gear selection is handled by a Gennie Shifter, connected to a TH400 automatic transmission. Engine compartment access is aided by designing the hood to lift on pneumatic cylinders, raising the bonnet straight up.

Down in the weeds is where the Snake lives, and this one has only 1/4-inch of ground clearance when the suspension is relaxed. It's hard to get any lower than that!

Bubble skirts were molded to the quarters, and trunk corners are rounded. Taillights from a 1979 Olds were deeply tunneled, as were front and rear license plates. Dual antennas sink into the left front fender. The windshield is flush-mounted, and the rest of the glass has been treated to tint and etching.

Custom work continued on the inside, with dash and windshield garnish moldings smoothly molded in. A Panasonic overhead console, filled with controls, was custom fit to the roof. Seats came from a 1976 Lincoln, and the steering wheel is out of a 1975 Corvette. The gas tank is a 1978 Corvette item, tucked in behind the rear seat.

Step back and take a look. This is one wild custom. And just about as low as a snake's belly.

The low profile was aided by a 5-1/2-inch top chop, and then a '50 rear window was installed. Bubble skirts, molded to the rear quarter, completely conceal the rear tires and wheels, adding to the low look.

You don't have to step high to get into this cruiser. It's more of a bend-down and slide into the seat operation. Interior comfort is assured by seats out of a '76 Lincoln, with door panel upholstery matching the theme.

Taillights are '79 Oldsmobile items, deeply tunneled. Trunk corners were rounded, the pan rolled and the bumper molded in.

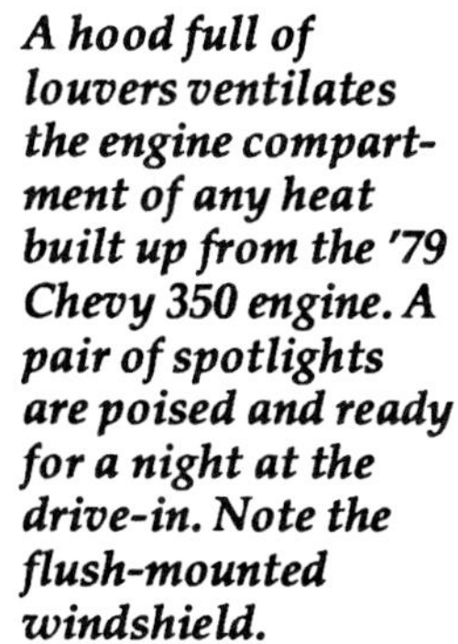

A hood full of louvers ventilates the engine compartment of any heat built up from the '79 Chevy 350 engine. A pair of spotlights are poised and ready for a night at the drive-in. Note the flush-mounted windshield.

Rectangular Ford headlights hide behind tinted Plexiglass, while '80 Mustang parking lights are tunneled in below. The grille is vintage '55 Oldsmobile, and the bumper has been molded to the rolled pan. Centered in the pan is a recessed license plate.

Below-Fuzzy dice hang from the rearview mirror on the overhead Panasonic console. This is a step into the high-tech future for a traditional style custom, but the execution worked perfectly.

by Rich Johnson

LOW & LOVELY

Vehicle: 1951 Merc
Owner: Ernie Roth — Selden, New York

Ernie Roth was just a hard-working man who had a dream. And like most working men, the dream was bigger than the pocketbook. Ernie's dream was to someday own a custom Merc.

Now, one thing you have to understand about Ernie is that even though he is a hard-working type of guy, he doesn't particularly enjoy working on cars. To him, the fun part isn't scraping his knuckles and living forever with grease under the fingernails. To Ernie, the fun part is turning the key and driving the car of his dreams.

But when the dream is to own a custom Merc, and the budget just won't accommodate it, the only answer is to build it yourself over a long period of time, a little here and a little there until it's finished. And that's exactly how Ernie approached it.

The year the dream started to come true was 1979. Ernie found the car in New Mexico, bought it and took it back home to New York. He drove the car around for a couple of years and then in 1981 started to go to work on it.

All the mechanicals were given the treatment, so the car would be reliable and drivable on a daily basis. The car was stripped to bare metal and Ernie started adding custom ideas he had been thinking about for a long while. By 1982, he was back on the road with the car, all finished. But as any hobbyist knows, a car like this is never really finished. As he went to more and more shows, Ernie discovered a lot of neat things about custom Mercs that he wanted to integrate on his own car.

So, back to the garage it went to have disc brakes, power steering, and several other goodies installed. This kind of work went on through the long winter months for a total of nine years. But that's exactly how a car like this is built on a budget.

Among the body mods that were done, the hood was nosed, peaked and corners rounded front and back. The trunk corners were rounded top and bottom. The gas door was filled and moved inside the trunk, the top was chopped 4 inches and all the windows are functional. Headlights were frenched and peaked, taillights sunken in 6 inches and '59 Cad taillights installed in '58 Chevy houses. Twin electric antennae were sunk in the door. A '55 DeSoto grille has taken its place in a contoured grille opening, and a '52 DeSoto bumper had the guards removed and filled. '53 Buick side trim was installed, and Appleton dummy spots increase the glitter. There are 125 hood louvers, Lakes pipes that are 80-inches long, bubble skirts and '56 Olds hubcaps.

The suspension was lowered 3-1/2 inches in front and 6-1/2 inches in the rear by using lowering blocks. Air shocks are used in the rear. 1971 Ford disc brakes and Ford power steering were installed.

On the inside, Ernie used black Naugahyde diamond tufted upholstery, hand-sewn, with matching material used for the side panels, headliner and trunk. The car is carpeted throughout.

Power comes from a 1971 Olds 350 with a pair of 4-barrel Carter AFBs and a turbo 400 transmission. A '71 Ford rearend was employed, with 2.75:1 gears. The engine was mounted 4 inches below the original engine position. Air cleaner and rocker arm covers are chromed, as are the hood supports, inner fenders, and radiator shield. All the air conditioner and cooling system lines are braided.

Whew! So, for a guy like Ernie, who doesn't like to work on cars, you can imagine how neat it felt to get to the end of all that work. Finally, he believes that the car is finished. And he's not alone in his belief. Along with winning numerous awards in a variety of categories at different shows around the country, Low & Lovely was chosen Best '49-'51 Merc at the James Dean show in Indiana in 1984. That's pretty impressive.

Ernie has just a few words he would like to pass along to all the backyard people who are working on their own dream cars. "Remember, you can do it. So don't give up. Keep building your dream and most of all keep cruisin'. That's the main thing."

Black Naugahyde upholstery was hand-sewn in a diamond tufted pattern. Matching material was used for side panels, headliner and trunk.

Power is derived from a '71 Oldsmobile 350 V8, breathing through a pair of 4-barrel Carter AFB carburetors. Braided lines and hoses add sparkle to the engine compartment, to say nothing of the extensive use of chrome.

Inside the upholstered trunk is a spare tire, covered with the quintessential nostalgic lyrics from popular music of the Fifties.

Radical custom work was performed up front. The headlights were frenched and peaked, for a wild look.

In the forward part of the passenger side door, a pair of electric antenna were sunk, a traditional custom trick that involves both form and function.

Appleton dummy spotlights are perched on each windshield pillar, and the hood has been ventilated with four rows of louvers, totaling 125 in all.

Top was chopped 4 inches, and rear quarter glass etched with a scene of cruising under the moonlight.

by Ron Ceridono
photos by Mike Meyer

FAMILY TRADITION

Vehicles: 1949 & 1951 Mercurys
Owners: Rod and Bob Beverly

Like a lot of young guys in the '50s, Bob Beverly was a custom car enthusiast. But as the responsibilities of being a husband and father grew, the time for pursuing his interest in customs seemed to vanish.

In 1983 Rod, Bob's son, convinced him to attend a custom car show. (The Beverlys obviously instilled in their son the proper values.) Three weeks after that show Rod found a '49 Mercury and started to customize it. Bob, with his interest in customs reawakened, jumped in to give his son a hand. It was only a matter of time. Soon Dad had his own project, a '51 Mercury convertible.

Rod's first attempt at customizing was ambitious to say the least, a 6-1/2 inch top chop his first time out. A '47 Buick grille replaced the original, headlights were frenched using '42 Merc rings.

The hood has rounded corners and has been nosed. Entry to the low-lid Merc is through solenoid operated doors. Lincoln taillights are separated by a smoothed off, electrically opened trunk lid. All the body work and painting was handled by 23 year-old Rod.

Inside, the sedan is still under construction. So far, a Cadillac steering column and power front seat have been installed. Rear seating is a '65 T-Bird wraparound unit.

Under the hood are 255 cubic inches of Henry's finest. The rebuilt '51 flathead features 3-2s, Offenhauser heads, and Mallory ignition. A 3-speed overdrive transmission and stock rearend complete the drivetrain.

Dad's car is more contemporary in the components department. A Camaro subframe and rearend have replaced the original suspension pieces. Power steering and brakes are benefits of the swap.

The convertible is powered by a Chevrolet 350 cid small block and automatic. Bob travels in comfort, as the Merc has been equipped with air, power windows, power seats, and a tilt and tele steering column.

Not to be outdone by his son, Bob whacked six inches out of the windshield of his custom. The Beverly clan takes top chopping seriously.

Making the top mechanism fold on such a severely chopped convertible turned out to be a major project in itself. Bob solved the problem by adapting modified Cadillac top irons to the Merc, and the top is completely operational.

Stock side trim has been removed and replaced by '51 Buick pieces. '52 DeSoto bumpers and a '48 Oldsmobile grille are used. Frenched Lincoln taillights bring up the rear.

When it came time to put color on the convertible, the Beverlys turned a 15'x25' shed into a paint booth. This transformation was accomplished by covering the interior walls and ceiling with plastic. Two vents, a 24" fan, and some fluorescent shop lights, and they were ready to paint. A homemade spray booth on the cheap. Rod, a first time painter, shot the Ditzler enamel. The results would make a pro painter happy.

The customizing tradition is alive and well in the Beverly family.

Both cars feature lakes pipes and Olds hubcaps. Lines of a chopped Merc are classic, even in examples as different as a sedan and a convertible.

Rod and Bob massage the body of the '51. Both cars were father and son efforts.

Windshield posts on the convertible were reinforced with steel rods running clear to the firewall.

Bob and our own Scott Smith flop the top on the '51.

A properly executed 6-inch chop of the convertible top resulted in a tight fit and low profile. It also left the top completely functional for open air driving. And, folded down, the top fits nicely beneath its boot.

The '49 gets its go from a decked out flathead. Polished aluminum Offenhauser heads and triple intake manifold make it look and run great.

The '51 has a small block Chevy under its hood, also with three twos. Power steering, brakes, and air conditioning are a part of the package.

A homemade paint booth was built inside a 15'x25' shed, covered inside by plastic sheeting. Rod's first-ever paint job did him proud.

Left-The convertible interior looks every bit as good as the exterior. Traditional style upholstery lends a nostalgic atmosphere, and the chopped glass lets you know this is one low ride.

Above-Profiles of the two cars have one thing in common, big time chops.

by Rich Johnson

PHIFTY 3

Vehicle: 1953 Mercury 2-Door
Owner: Bill Tripp — Tulsa, Oklahoma

Interior decor was dressed up with '55 Mercury Montclair material and a bit of Bill's own imagination. This is a mild custom that retains such stock components as door handles and crank window regulators, but incorporates all the right touches of the Fifties custom period.

When a fellow stopped by Route 66 Detail & Vintage Parts to admire Bill Tripp's Mercury, he looked it over from bumper to bumper and then declared, "I had one exactly like this in 1952." Bill just smiled and thanked the visitor for his compliments. The man's intentions were good, even though his judgement was in error. Nobody has ever had a car exactly like Bill's.

Actually, when people first see this '53, they tend to think it's a '54 because of the wrap-around taillights. Then, as they look farther, they see that something else was done, then something else. That's the way it is with well executed mild customs.

To the casual observer, the custom tricks become gradually apparent. A fifth tooth was added to the grille. Headlights were frenched, lakes pipes installed, and Lancer wheel covers added. The radio antenna has been tunneled, a Batman-style scoop installed on the hood, and a custom upholstery job done. There's more, and all of it's done with such craftsmanship that some folks are fooled into thinking it came this way from the factory.

But let's take a look beneath the surface. That's where the real trick stuff has been done. Bill found this car sitting outside, being abused by weather, tree sap, and birds. He intended to put the car back into roadworthy condition, but wanted better ride and handling characteristics than would be delivered by simply upgrading stock components. So, he decided to swap the body onto a late-model chassis.

A friend, Art Maimbourg, had done a chassis swap on his '53 Henry J, and everything worked out just fine, so Art told Bill that he would help out with this Mercury project. Unlike some frame swaps, Bill decided to use his '67 Chevy Impala as a chassis donor because it had always been dependable, it had only 67,000 miles on it, it had been in his family since new, it got good mileage (25 mpg), and it had most of the amenities that Bill wanted. The decision had nothing to do with dimensional compatibility for the swap, or anything as technical as that. He simply liked the Chevy, and he wanted it under his Merc. Anything can be made to fit, right?

As it turns out, there are certain similarities between a '67 Impala and a '53 Mercury. For instance, factory weight for the two cars was within about 10 pounds, give or take a little fudge factor. Wheelbase for the Impala is 119" and for the Merc it is 118". Close enough. Overall length is about 11" different, but that could be adjusted. It was in the area of front and rear

With the '53 Merc body swapped onto a '67 Chevy Impala chassis, Bill got it all — late model steering, brakes, and suspension along with classic styling of an earlier era. Note where the firewall was cut and matched, then rewelded. Enough of the Impala firewall was retained to allow Chevy pedals and electrical junction boxes to remain.

The Merc floorpan was used under the back seat, to retain the stock seat mounting tabs. Then the stock trunk was rewelded in, due to sender accessibility and filler neck position. Handmade sheetmetal taillight bezels bolt in place.

One custom touch is the tunneled antenna on the passenger side rear fender. Here is a photo of the hole that was cut and rolled around the lip in preparation to building the recess.

Headlights were frenched, and a Batman-style scoop built into the leading edge of the hood. Note also the Dagmars, prominently positioned on the front bumper.

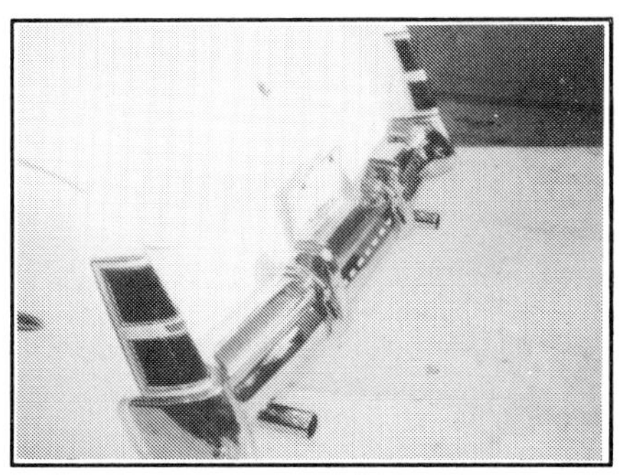

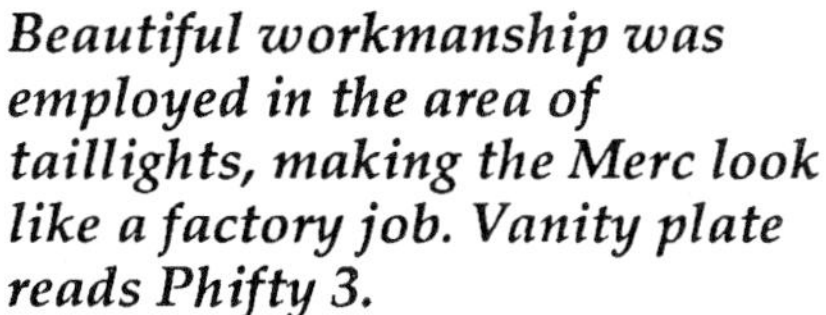

Beautiful workmanship was employed in the area of taillights, making the Merc look like a factory job. Vanity plate reads Phifty 3.

tread width that things got interesting. Chevy had a 62.5" front tread width, compared with the 58" dimension for the Merc. Rear measurement for the Chevy is 62.4" and a relatively narrow 56" for the Merc. Hmmm. Something would have to be done.

After some preliminary measuring and removal of everything that could burn, Art began systematically cutting the Chevy body from its floor. The intent was to retain the Chevy toe board and main firewall, so the pedals and electrical junction boxes would still fit. Both firewalls were measured, then cut in such a way as to fit together when the Merc body was lowered over the Chevy chassis. Welding was done at the firewall and along the floor just above the frame adjacent to the door sill plates.

Of course, the Chevy frame was just a bit too wide, so Art cut the center of the floor pan, front to rear, and widened the body from the greenhouse down just enough to slip over the frame and give adequate clearance to the tires, then re-welded the floor. Slick as a whistle.

The Merc floor pan was used under the back seat, to maintain the stock seat mounting tabs. A 1972 Cadillac seat frame was grafted to the Merc seat frame to fit these tabs. It's a tight squeeze, but the seat does fit. A support for the rear seat was fabricated out of round stock, shaped to match the contour of the seat cushion, and braced from the driveshaft tunnel. Interior decor was upgraded with '55 Montclair upholstery and a bit of imagination. The Merc trunk floor was retained, due to the sender accessibility and filler neck in the center of the body. A new tank was fabricated to help solve clearance problems with the anti-sway bar.

Some of the inner fender panels had to be trimmed, and the suspension was modified to give a more pleasing ride quality. Rear coils were shortened by one loop. The anti-sway bar was retained, to keep the body off the tires. The fender lips were rolled under and stock 14" wheels and P195/75R14 tires employed. This kept the body low and gave the necessary clearances.

This is one of those cars that will probably never be totally finished. A continental kit is planned for the future, and continual inspiration comes from Bill's study of other cars. But that's just the way it is with a mild custom.

Sources

Antique Auto Glass Service
6108 Kentucky
Raytown, MO 64133
(816) 887-2462

Birdhaven Vintage Auto Supply
RR1, Box 153
Colfax, IA 50054
(515) 674-3949

Bob Drake Reproductions
1819 NW Washington Blvd.
Grants Pass, OR 97526
(503) 474-0043

Bradley Antique Automotive
4200 South Interstate 85
Charlotte, NC 28214
(704) 392-3206

Brockman Mellow Tone
940 S. Washington St.
Hagerstown, IN 47346
(317) 489-4111

Butch's Rod Shop
2853 Northlawn Ave.
Dayton, OH 45439
(513) 298-COOL

C & G Early Ford Parts
165 Balboa
San Marcos, CA 92065
(619) 744-0470

Concours Parts & Accessories
3563 Numancia St.
Santa Ynez, CA 93460
((805) 688-7795

Dennis Carpenter Ford Reporductions
Box 26398
Charlotte, NC 28221
(704) 786-8139

Dick Dean
625 West Valencia
Fullerton, CA 92632
(714) 879-3231

Early Ford Parts
2984 Summer Ave.
Memphis, TN 38112
(901) 323-2179

ECi
Vernon, CT
(203) 872-7046

Fat Man Fabrications
8621-C Fairview Road, Highway 218
Charlotte, NC 28227
(704) 545-0369

Fender Skirts Co.
1801 Jason Street
San Diego, CA 92154
(619) 424-9007

Ford Flathead Shop
225 North Commerce Street
Gillman, IL 60938
(815) 265-4375

Gene Winfield Rod & Custom Contruction, Inc.
7256 Eton St.
Canoga Park, CA 91303
(818) 883-2611

Hot Rod & Custom Supply
620 NE 15th Ave.
Cape Coral, FL 33909
(813) 574-7744

J & M Enterprises
Box 510023
St. Louis, MO 63151
(314) 846-7057

LeBaron Bonney Company
6 Chestnut St.
Amesbury, MA 01913
(508) 388-3811

McDonald Ford Parts
RR 3, Box 94
Rockport, IN 47635
(812) 359-4965

Mercury And More
29007 Northeast 88th Ave.
Battle Ground, WA 98604

Mercury Research Company
639 Glankler St.
Memphis, TN 38112

Merle Fourez
1393 Shippee Lane
Ojai, CA 93023

Mr. Merc
5980 Oneka Lake Blvd.
Hugo, MN 55038
(612) 426-5128

Nertz's Neat Car Stuff
945 Parker Avenue N.W.
Medicine Hat, Alberta
Canada T1A 6W7
(403) 527-9482

Night Prowlers
Box 551
Lamar, MO 64759
(417) 535-6771

P & J Automotive
6262 Riverside Dr.
Danville, VA 24541
(804) 822-2211

Papke Enterprises, Inc.
17202 Gothard, Unit 3
Huntington Beach, CA 92647
(714) 839-3050

Pat Walsh Auto Parts
Box Q
Wakefield, MA 01880

Racecars In Retrospect
Box 3116
Alliance, OH 44601
(216) 823-6167

Radio Restoration - Antique Automobile Radio
Box 892
Crystal Beach, FL 34681
(813) 785-8733

Rhode Island Wiring Service
Box 3737
Peace Dale, RI 02883
(401) 789-1955

Shoebox Ford Parts
9319 Southeast 29th Street
Midwest City, OK 73130
(405) 732-6027

Speedway Motors
Box 81906
Lincoln, NE 68501
(402) 474-4411

Thunder Road Motor Works, Inc.
RD2, Box 8
Hudson, NY 12534
(518) 828-3280